Gardens of Prehistory

Gardens of Prehistory
THE ARCHAEOLOGY OF SETTLEMENT AGRICULTURE
IN GREATER MESOAMERICA

Edited by Thomas W. Killion

The University of Alabama Press
Tuscaloosa & London

Copyright © 1992
The University of Alabama Press
Tuscaloosa, Alabama 35487–0380
All rights reserved
Manufactured in the United States of America

designed by zig zeigler

The paper on which this book is printed meets the minimum requirements of American National Standard for Information Science-Permanence of Paper for Printed Library Materials, ANSI Z39.48-1984.

Library of Congress Cataloging-in-Publication Data

Gardens of prehistory : the archaeology of settlement agriculture in Greater Mesoamerica / edited by Thomas W. Killion.
    p.   cm.
Edited papers from a symposium held May 9, 1987, in Toronto.
Includes bibliographical references and index.
ISBN 0-8173-0565-3 (alk. paper)
    1. Indians—Agriculture—Congresses. 2. Agriculture, Prehistoric—America—Congresses. 3. Land settlement patterns, Prehistoric—America—Congresses. 4. Indians—Antiquities—Congresses.
5. America—Antiquities—Congresses. I. Killion, Thomas W.
E59.A35G37   1992
630'.972'0901—dc20   92-3420

British Library Cataloguing-in-Publication Data available

Contents

Illustrations   ix

Preface   xiii

1
The Archaeology of Settlement Agriculture
   THOMAS W. KILLION   1

2
Factors Affecting Settlement Agriculture in the Ethnographic and Historic Record of Mesoamerica
   WILLIAM T. SANDERS AND THOMAS W. KILLION   14

PART I: Settlement and Agriculture in the Arid Lands of Greater Mesoamerica   33

3
The Southwestern Ethnographic Record and Prehistoric Agricultural Diversity
   TIMOTHY D. MAXWELL AND KURT F. ANSCHUETZ   35

4

House-Lot Gardens in the Gran Chichimeca: Ethnographic Cause for Archaeological Concern
  WILLIAM E. DOOLITTLE   69

5

The Productivity of Maguey Terrace Agriculture in Central Mexico During the Aztec Period
  SUSAN T. EVANS   92

PART II: Artifact Distributions and the Organization of Prehistoric Agriculture: Evidence from Lowland Mesoamerica   117

6

Residential Ethnoarchaeology and Ancient Site Structure: Contemporary Farming and Prehistoric Settlement Agriculture at Matacapan, Veracruz, Mexico
  THOMAS W. KILLION   119

7

A Consideration of the Olmec Phenomenon in the Tuxtlas: Early Formative Settlement Pattern, Land Use, and Refuse Disposal at Matacapan, Veracruz, Mexico
  ROBERT S. SANTLEY   150

8

Agricultural Tasks and Tools: Patterns of Stone Tool Discard Near Prehistoric Maya Residences Bordering Pulltrouser Swamp, Belize
  PATRICIA A. MCANANY   184

CONTENTS

PART III: Prehistoric Cultivation, Landscape Modification, and Chemical Characterization   215

9
Intensive Raised-Field Agriculture in a Posteruption Environment, El Salvador
CHRISTIAN J. ZIER   217

10
Prehistoric Intrasettlement Land Use and Residual Soil Phosphate Levels in the Upper Belize Valley, Central America
JOSEPH W. BALL AND RICHALENE G. KELSAY   234

PART IV: Summary and Critique
B. L. TURNER II AND WILLIAM T. SANDERS   263

References   285

Contributors   325

Index   330

Illustrations

FIGURES

P-1.	Geographical Loci of Volume Chapters	xv
3-1.	Schematic of Hopi Village and Environs	40
3-2.	Schematic of a Hopi Garden Complex	43
3-3.	Hopi *Akchin* Field and Common Facilities	48
3-4.	Prehistoric Sites in Lower Rio Chama Valley	55
3-5.	LA 48679; Study Unit 6	62
3-6.	LA 48679; Study Unit 13	63
4-1.	Map of the Gran Chichimeca	71
4-2.	Maize in a House-Lot Garden	75
4-3.	House-Lot Garden	80
4-4.	Abandoned Garden Terraces	81
4-5.	Rock Ring at Site Son K:4:20 OU	84
4-6.	Parallel Rock Alignments at Site Son K:4:46 OU	86
4-7.	Rock Wall at Site Son K:4:118 OU	87
4-8.	Rock Enclosure at Site Son K:4:48 OU	88

ILLUSTRATIONS

4-9.	Low Rock Enclosure at Site Son K:4:144 OU	89
4-10.	Composite Map Showing Stylized Mesa Top	90
5-1.	Central Highlands of Mexico	94
5-2.	Teotihuacan Region Settlement Pattern	96
5-3.	Cerro San Lucas and the Village of Cihuatecpan	98
6-1.	The Tuxtlas Region and Matacapan	124
6-2.	Schematic Representation of Spatial Components of House-Lot Model	125
6-3.	Simplified Plan of Contemporary House Lot, Northern Tuxtlas Region	133
6-4.	Simplified Plan of Contemporary House Lot, Southern Tuxtlas Region	134
6-5.	Topography and Mounded Architecture at Matacapan	138
6-6.	Ceramic Distributions and Habitational Mounds, Area I	141
6-7.	Ceramic Distributions and Habitational Mounds, Area II	142
7-1.	The Western Tuxtlas	151
7-2.	Formative Period Sites in Matacapan Area	156
7-3.	Histogram of Pottery from Off-Site Squares	160
7-4.	Histogram of Pottery Stratified by Spatial Zone	162
7-5.	Histogram of Vessel Forms	165
7-6.	Topographic Map of the Teotihuacan Barrio	170
7-7.	Early Formative Refuse Densities, Teotihuacan Barrio	174
8-1.	Belize, Central America	186
8-2.	Shipibo Agricultural Tasks by Field Type	191
8-3.	Kapauku Agricultural Tasks by Field Type	192

ILLUSTRATIONS

8-4.	Settlement Bordering Pulltrouser Swamp	199
8-5.	Changes in Oval Biface and Debitage	203
8-6.	Patterns of Oval Biface Breakage	206
9-1.	Map of Central and Western El Salvador	220
9-2.	Artificial Cut Exposing Cerén Site House	223
9-3.	Overhead View of Test Pit of Cornfield	225
9-4.	Cast of Young Maize Leaf	226
9-5.	Salvadoran Farmers with Digging Sticks	229
10-1.	Buenavista de Cayo-Guerra Study Area	236
10-2.	Guerra Site	238
10-3.	Patio Group 4-C-1, Guerra	246
10-4.	Soil Phosphate Levels, Transect B, Guerra	251
10-5.	Soil Phosphate Levels, Transect I, Guerra	253
10-6.	Buenavista del Cayo Site	256
10-7.	Soil Phosphate Levels, Transect B, Buenavista	257

TABLES

3-1.	Hopi Garden Acreage	41
3-2.	Hopi Garden Crops	42
3-3.	Prehistoric Pueblos in the Lower Rio Chama	56
5-1.	Land and Landholding	101
5-2.	Area and Productivity of Agricultural Land, Cerro San Lucas	108
6-1.	Percentage of House-Lot Area Occupied by Use Areas in Portions of the Tuxtlas Sample	130
6-2.	Correlation Matrix of Site Structural Components and Factors of Agricultural Production	135

6-3. Areal Percentage Comparison of House-Lot Model Components and Prehistoric Ceramic Density Distribution Quartiles — 145

8-1. Biface Tool Fragments from Residential Middens and Interplatform Contexts — 208

8-2. Debitage from Platform Middens and Interplatform Contexts — 210

Preface

THE PREHISTORIC agricultural systems of the New World provided the foundations for a diverse set of complex social developments ranging from the Puebloan societies of the American Southwest to the archaic state polities of Mesoamerica and the Andean region. From the tropical forests of Central America to the high arid environments of northern New Mexico, native American farmers made use of a distinctive set of cultigens and cropping systems that supported, with varying degrees of success, growing populations and expanding economies. Lacking most domesticated animals, so important to the mixed agricultural systems of the Old World, Precolumbian farmers developed intensive and resilient systems of agricultural production. These systems supported societies that, to varying degrees, altered the landscapes they inhabited and generated an archaeological record that chronicles the evolution of farming in the New World.

SCOPE

Gardens of Prehistory: The Archaeology of Settlement Agriculture in Greater Mesoamerica consists of a group of studies directed at understanding the organization and structure of prehistoric agriculture in archaeological perspective. Areally, the volume covers recent research in northern New Mexico, the United States; Sonora and southern Veracruz, Mexico; and Belize and El Salvador, Central America (fig. P-1). The studies focus topically on the identification of cultivation practices implemented by prehistoric agriculturalists within or close to the farming settlement, generally referred to as *house-lot gardens* and *infields*. The research efforts presented in this volume are directed at using the archaeological record to diagnose the nature of agricultural production from kitchen gardens located within the residential lot to staple-producing plots, or *outfields*, located at greater distances from ancient settlement zones.

Identifying the structure and organization of prehistoric agriculture has always represented one of the most difficult fields of inquiry for the archaeologist. The studies included in *Gardens of Prehistory* present innovative methodological approaches to the identification of ancient agricultural systems from a diverse set of New World environments. In doing so they not only broaden our understanding of native American food production systems but also strengthen our framework for inferences concerning the role of agriculture in the evolution of complex society.

CONTENTS

This volume is the outgrowth of a symposium on prehistoric gardens and infield agricultural systems held at the fifty-second annual meeting of the Society for American Archaeology held in Toronto, Canada, May 9, 1987. The edited collection is divided into four parts, which follow the preface, an introductory chapter, and a chapter on ethnographic and historic examples of settlement agriculture from lowland and highland Mesoamerica. Part I begins the analytical core of the volume and focuses on three research projects, one from the southwestern portion of the United States, another from northwestern Mexico, and a third from the highlands of central Mexico. All three studies stress the importance of residential gardens and infields to the overall spatial organization of farming around prehistoric settlements in the arid or more temperate environments of central highland Mesoamerica

PREFACE

Figure P-1. Geographical Loci of Volume Chapters

and its northern periphery. The studies presented on northern Mexico and the southwestern United States, while technically reporting on a temperate zone of North America, are included here because the agricultural systems of these arid landscapes were based on cultigens imported from the tropical lands that are the focus of the volume. The research presented in Part I indicates higher levels of agricultural intensity and a more diverse set of horticultural and agricultural adaptations than are generally assumed for this region.

In the second part of *Gardens of Prehistory* the focus shifts to the humid tropical environments of lowland Mesoamerica. Part II presents

xv

three new studies that compare ethnoarchaeological and ethnographic models of peasant farming behavior with recently recovered archaeological data bases. These studies demonstrate how an understanding of contemporary subsistence farming and refuse disposal behavior can be used to characterize the systematic use of settlement and agricultural space by ancient cultivators and to explain the patterned distribution of ceramic and lithic artifacts that are found in and around ancient communities. Case studies come from the Gulf Coast lowlands of southern Veracruz (the Olmec heartland) and the lowland Maya region of northern Belize.

Part III of the volume deals with chemical signatures and landscape modification as keys to understanding the structure and organization of settlement agriculture in prehistoric America. The first study examines the fortuitous preservation of prehistoric plants and planting surfaces adjacent to residences under a volcanic ash deposit in the Zapotitán Valley of El Salvador. The last analytical chapter in Part III uses phosphate analysis on a larger spatial scale to examine the agricultural and other uses of intrasettlement space among a group of Classic period Maya sites in central Belize.

The final section of *Gardens of Prehistory* is a summary and critique chapter consisting of comments by B. L. Turner II and William T. Sanders, discussants at the original symposium in Toronto. In this final chapter the authors, both of whom have made important contributions to the study of prehistoric agricultural systems in the Americas, set forth some of their recent thoughts on the study of prehispanic agriculture. Following some general observations on the present state of prehistoric agricultural studies, they take a critical look at the chapters in the volume, identifying the advantages and disadvantages of each of the approaches and setting forth some important methodological and theoretical goals for the future.

CONTRIBUTION

In sum, *Gardens of Prehistory* contains eight chapters presenting the results of recent research on settlement agriculture (infield and residential gardening) as practiced by a sample of prehistoric societies in the

PREFACE

New World. In addition to presenting basic data, each contribution (1) develops new analytical procedures for the identification of ancient agriculture, (2) formalizes a theoretical model of agricultural practices conducted in and around settlements, and (3) defines the spatial structure of settlement as perceived by the archaeologist. Introductory and concluding chapters emphasize the overall theme of the volume (the effect of gardening and infield agriculture on the use of space by farming communities and the effect of this process on the archaeological record) and identify problems for future research.

The primary contribution of the volume is the presentation of a set of new and innovative methods directed at identifying the structure and organization of ancient agricultural systems. Equally important, though, are the volume's emphases on the diversity of traditional residential and infield agriculture practices, the contribution of infield agriculture to subsistence, and the overall effect of this agricultural regime on the spatial structure of prehistoric settlement.

Gardens of Prehistory fills a present void in the selection of volumes concerned with the origins and development of prehistoric agricultural systems and could prove to be a valuable adjunct to upper-level undergraduate and graduate courses covering these topics. This collection of papers will be of particular interest to anthropologists and archaeologists concerned with the structure, organization, and identification of prehistoric agricultural systems in the New World and throughout the tropics. The volume offers new insight into our understanding of native American agriculture and the effect of varying land-use techniques on the form of ancient settlements. This volume will also attract a wider audience of geographers and agronomists interested in traditional agricultural practices in arid/highland and humid/tropical environments.

CHAPTER 1   THOMAS W. KILLION

The Archaeology of Settlement Agriculture

THE CULTIVATION OF CROPS within and surrounding settlements was a fundamental feature of everyday life among many of the prehistoric farming populations of the New World. Alone or in conjunction with outfield production on fields at greater distances from the residence, infield and house-lot crops provided staple caloric support and important nutritional supplements and served a host of other household economic needs. Agricultural practices carried out in proximity to residential areas conditioned the use of space, the layout of structures, and the deposition of debris and other residues that have become part of the archaeological record.

    Such evidence represents the material fallout of one of the most intensively utilized portions of the prehistoric landscape. Our notions concerning the nature of agricultural production in near-residential and infield contexts, subsumed here under the heading of settlement agriculture, influences the analysis of and meaning we attribute to the distribution of ceramic and lithic debris, the remains of prehistoric buildings and features, and the chemical and other physical signatures of ancient habitation and agriculture. The theoretical importance of settlement agriculture to early food production systems also has become clearer as scholars begin to examine the spatial significance and energetic contribution of near-residential and infield agriculture to prehistoric cultivation systems.

Gardens of Prehistory: The Archaeology of Settlement Agriculture in Greater Mesoamerica examines archaeological research on prehistoric agriculture within the near-residential and infield contexts directly adjacent to settlements from a sample of tropical and subtropical regions in prehispanic America (see fig. P-1 in Preface). The volume focuses on examples of settlement agriculture from the North American Southwest and from Mesoamerica and its southern periphery. The goal of the volume is to examine settlement agriculture from an archaeological and analytical perspective. This approach underscores the importance of near-residential and infield cultivation to the investigation and interpretation of one of the most common contexts of archaeological survey and excavation and contributes to the establishment of more secure methods for reconstructing prehistoric settlement and agricultural systems in such contexts.

ARCHAEOLOGICAL APPROACHES TO THE STUDY OF PREHISTORIC AGRICULTURAL SYSTEMS

The practice of agriculture on areas within and adjacent to settlements is just one of a growing number of agrotechnologies now recognized by archaeologists as fundamental components of prehistoric New World agricultural systems. It is now generally acknowledged that the complex societies of prehistoric tropical and subtropical America practiced a diverse set of extensive and intensive agricultural technologies capable of sustaining relatively dense populations in many regions. Until quite recently, however, the use of areas in proximity to settlements had been largely undefined or characterized as a homogeneous zone of extensive, shifting cultivation in general syntheses of prehistoric agriculture (Grigg 1974:21–22).[1] Among Mesoamerican specialists this extensive model presented a somewhat anomalous appendage to the otherwise impressive achievements of prehistoric complex societies throughout the New World. While exceptions were put forward to the extensive model, few of these alternative models have been tested against the archaeological record in a systematic fashion.

ARCHAEOLOGY OF SETTLEMENT AGRICULTURE

Prehistoric irrigation- and terraced-based agricultural systems, notable exceptions to the extensive perspective, long provided the basic model of agricultural production supporting "high culture" in certain portions of the Andes and highland Mesoamerica. More intensive wetland and humid-based agricultural systems, however, have recently begun to figure prominently in the reconstruction of prehistoric agriculture and settlement in many other regions. Raised-field production in the eastern lowlands of Mesoamerica and the intensive cultivation of river levee lands along the Gulf Coast now appear to have been relatively common adjuncts to systems of settlement in many lowland areas of greater Mesoamerica, and raised fields, along with terraced-based agriculture, are now recognized as basic elements of prehistoric Andean systems of cultivation. Intensive agricultural and horticultural practices including the construction of gravel-mulched gardens, small-scale irrigation systems, microtopographic slope management, and other forms of landscape modification within and around settlements are also now regularly identified in subtropical and temperate regions north of Mesoamerica, suggesting sustained agricultural production and adaptation to environmental risks by prehistoric farming populations on either side of the Rio Grande.

Ubiquitous evidence of prehistoric settlement agriculture, however, has proven more resistant to archaeological identification and treatment. In the absence of relict agricultural features and concrete evidence of landscape modification, prehistoric agricultural production in proximity to residential or settlement areas has been generally posited on the basis of analogy to contemporary and ethnohistorically known agricultural systems. Archaeological research conducted within the spatial context of settlement agriculture and keyed to the potential of the overall agroecosystem provides a means of augmenting and enhancing the recovery of data pertinent to our understanding of this critical component of prehistoric agricultural production. In the sections that follow I examine the importance of use of space within and adjacent to farming settlements for agricultural production and the implications of this use to the formation of the archaeological record as presented in the chapters constituting this volume.

SPATIAL AND MATERIAL PROPERTIES OF SMALL-SCALE AGRICULTURAL SYSTEMS

While geographers, ethnographers, and archaeologists have long been aware of the effect of agriculture on settlement patterns, the spatial juxtaposition of farming operations and residence has rarely had much of an impact on the collection and analysis of archaeological data in these contexts. The spatial organization of agricultural production has profound implications for the physical properties of small-scale farming societies, not only affecting the positioning of structures and facilities but also altering the regular deposition of domestic debris and the resulting distributional properties of the archaeological record. As Sanders and Turner note in the concluding chapter of this volume, the spatial form of agricultural production is also linked to systems of land tenure and other nonagricultural factors associated with the growth and organization of complex society.

Among nonindustrial farming societies in both humid and arid tropical environments, the residential areas within settlements and the zones surrounding them serve a variety of purposes. These areas provide space for household production, food consumption, shelter, and storage and places for social interaction, recreation, and leisure. Tropical as well as temperate settlements and their immediate environs also exhibit a wide range of agricultural uses, and areas located in and around settlements are often fundamental and highly productive components of the overall agroecosystem. The material pattern resulting from these uses reflects a basic fact of life in the tropics—that much cultivated space is also heavily used habitational space and as such bears a strong material trace of both cultivation and residential living.

Archaeologists direct the majority of their observations and fieldwork efforts to those portions of the landscape that contain the richest habitational remains—the locations of ancient settlements. As noted above, settlements can play host to a mixture of habitational and productive (including agricultural) uses. Alternatively, settlement and agricultural systems exist where the spatial overlap between residence and field is less pronounced, and the consequence for both the organization of production and the archaeological record is a reduction in the amount of material overlap. In both cases the structure of the archae-

ARCHAEOLOGY OF SETTLEMENT AGRICULTURE

ological record is greatly affected by the organization of agricultural production in and around settlements. An important problem addressed in the present volume is the development of methods that permit the identification of cultivated and residential spaces and the intensity of agricultural production on the lands closest to settlement.

The identification of agricultural and other uses for settlement and near-settlement space rests on an understanding of the effect of cultivation intensity and land use within these contexts. Along with accepting higher population densities for some of the prehistoric farming cultures of tropical and subtropical America, archaeologists have come to realize that many settlement and subsistence systems were more complexly organized than a simple model of low-intensity shifting cultivation would suggest. Subsistence farmers make use of a wide range of crops, cultivation techniques, and fallowing regimes in these environments today in order to reduce environmental risks and ensure sustained yields. It is therefore necessary to examine the possibility that a similar range of agricultural practices was employed in the past in order to provide sufficient food supplies for prehistoric populations. It is also recognized that postoccupational formation processes heavily influence the pattern of remains examined by the archaeologist.

Under conditions of low population density, shifting cultivators are able to expand into virgin territory by moving settlements and fields whenever nearby agricultural land becomes exhausted. As population levels increase and the landscape becomes more crowded, the alternative of migration to uninhabited regions is constrained and the problem of continuous agricultural production must be solved within smaller and more tightly bounded territories. Both temperate and tropical cultivators often solve this problem by stratifying their available territory into zones of differential use. Agricultural plots located close to permanent settlements are more easily worked than more remote plots, hence land within or directly adjacent to residential areas is often more intensively farmed than land along the periphery of a community's territory.

A model of concentric zones of land use around settlements with successive rings utilized in a more extensive fashion has been employed to describe many systems of subsistence agriculture in the tropics (Beckerman 1983; Hiraoka 1986; Moran 1979; Ruthenberg 1971). In temperate Europe a similar model, known as the infield-outfield division of farms, was employed some time ago to distinguish manured and continuously cultivated fields adjacent to towns from more distant and longer fallowed parcels (e.g., McCourt 1955). The infield-outfield dichotomy can also be applied to tropical and subtropical contexts where cultivators who settle on prime agricultural land often employ short fallow (intensive) agricultural regimes on infields close to settlement and long fallow (extensive) regimes on more remote, and often marginal, lands.

Throughout the tropics, subsistence farmers also make use of the immediate residential zone or house lot for smaller, generally multipurpose, garden plots that are naturally fertilized by the continuous deposition of food waste, excrement, and other debris produced by household members and dooryard animals.[2] Generalized infield-outfield systems have been described for a number of contemporary and hypothesized prehistoric, New World contexts (Killion 1987a; Palerm and Wolf 1961; Sanders 1981; Vogt 1969).

While it is still unclear what absolute levels of population density were achieved prehistorically in the New World, it is nonetheless apparent that even moderately high densities could have had important consequences for both settlement and patterns of land use. With an increase in the number of settlements per unit of land and a reduction in the overall size of individual territories, outfield practices at the periphery of a settlement's territory could have been eliminated or greatly de-emphasized. As populations reached progressively higher levels within circumscribed territories, the practice of compartmentalizing available land into areas of differentiated agricultural use may not have provided sufficient food supplies, and more uniformly intensive systems of production would have replaced the infield-outfield structure.[3]

These conditions put pressure on the infield component of the agricultural system that may have required further capital investment (i.e., physical transformation of the field environment) or the redefinition of settlement and agriculture through the spatial reorganization of residence and field location. In the first case, irrigation, field raising, or terracing may replace field-fallow systems. On the other hand, where prime agricultural land is the focus of infield production, settlement may shift from a nucleated residential node at the core of an infield-outfield system to a more dispersed pattern of individual or small clusters of residences spread out across a landscape transformed into an intensively cultivated infield-residential mix. This process was observed in southern Nigeria, for example, where Ibo settlement changed from more nucleated town plans to a more dispersed compound pattern with the elimination of outfield agriculture under the pressure of growing population and increasingly specialized agricultural production (Buchanan and Pugh 1955; Netting 1977; Udo 1965). Distant bush-fallowed farmlands (outfields) were completely replaced by household-manured compoundlands (garden/infields) devoted to palm oil production as a completely overlapping pattern of settlement and intensive cultivation evolved (Udo 1965:fig. 8).

Recent discussion of nucleated and dispersed patterns of prehistoric settlement in Mesoamerica has touched on the organization of agricultural labor and the location of residence as processes related to the growth of population and the intensification of agricultural production (Drennan 1988; Farriss 1984; Sanders 1957, 1981). Problems of transport and the maintenance of soil fertility in the absence of large domesticated animals are seen as factors favoring settlement dispersal during periods of peak population growth. Farming households would have been most able to increase labor input and intensify agricultural production as the distance between field and settlement was reduced. While the implications of these factors have been examined at the level of settlement pattern analysis, much less attention has been given to their relationship to the distribution of artifactual debris, landscape modification, and soil alteration in and around settlement.

A basic characteristic of any agricultural system is the distance between the agricultural field and the residence. Generally farmers locate their most labor-demanding sites of cultivation as close to the residence as possible in an attempt to minimize labor expenditures. The differential use of nearby and more remote locations requires different groups of people, provisioned and organized for different sets of tasks. The cultivation of house-lot garden crops, for example, might best be organized as small but continuous inputs of labor by household members while the cultivation of a distant field crop could best be managed with a few visits by a team of laborers when key elements of the cultivation cycle demand large labor inputs. These organizational factors give rise to a zone of intensive, staple-producing agriculture surrounding settlements and, if available, an outlying area surrounding this zone reserved for less labor-intensive cultivation, hunting and gathering, and other resource procurement activities. Most nonindustrial systems of simple agricultural production exhibit a mixture of nearby intensive cultivation practices and more remote extensive strategies. The activities of people living and working in such a system can be envisioned as taking place within three generalized areas: (1) the residential lot, (2) the area outside the lot but within or adjacent to settlement, and (3) areas located at greater distances from the residence but outside the zone of land directly adjacent to the settlement. House, garden, and field areas thus constitute a spatial model of settlement and agriculture based on the location of people and the treatment of their territorial holdings from the walls of the dwelling outward.

A focus on cultivation adjacent to the residence probably represented one of the most common and longstanding means of intensification available to the densely populated farming societies that began to flourish throughout tropical America by the first millennium B.C. "Dooryard" (house-lot) gardens and nearby infields, long posited as important contexts in the early domestication of plants, represented key features of the later prehistoric urban landscapes among many New World societies playing a critical role in the overall system of agricultural production.

PROBLEMS OF DATA, METHOD, AND THEORY

It is a basic truism that dominant theoretical orientations have a direct effect on the collection and analysis of data. A move away from a narrow conception of prehistoric agriculture in the Americas as a homogeneous and redundant mix of a small number of crops, raised using simple techniques of shifting cultivation and having limited productive potential, had a definite effect on the quality and quantity of data pertinent to agriculture recovered from the archaeological record. New data and broadened expectations for what these data can reveal about the past have correctly stimulated the theoretical treatment of prehistoric agriculture but have also overshadowed problems of method and data recovery, obscuring the relationship between the facts of the archaeological record and the goal of understanding human organization and culture change. It is still the case that most information relating to the ancient agricultural societies of the New World (generally ceramic and lithic data from excavations and surface collections) is only weakly linked, in any analytical sense, to the ascendant paradigm of high productive potentials, diverse technologies, and organizational variability now dominating the investigation of prehistoric agriculture.

One of the most common methods for reconstructing ancient agriculture has been to turn to settlement pattern data to generate arguments concerning land-use strategies and changes in these strategies over time. Although settlement pattern studies have shown that strong correlations exist between particular classes of land and ancient settlement locations, they cannot tell us about the kinds of crops grown or the intensity of cultivation at those locations, that is, the system of cultivation employed within and around settlements. In the final analysis, the reconstruction of prehistoric agricultural systems is largely dependent upon our success at understanding the structure of the archaeological record and bringing both conventional data bases (ceramics and lithics) and less conventional (feature, landform, and chemical) analyses to bear on the spatial organization of settlement agriculture and its material outcome. For the time being archaeologists must still work toward the development of reliable methods that will help close the gap between data and the range of theoretical possibilities.

Gardens of Prehistory: The Archaeology of Settlement Agriculture in Greater Mesoamerica is an attempt to narrow the middle-range gulf facing the study of prehistoric agriculture by presenting one set of possible approaches to understanding agricultural production within or close to the farmer's residence. As such, the studies are essentially spatial and examine a scale of analysis relevant to the settlement and its immediate environs. This midrange scale of analysis, a generally less developed sector of archaeological research, fits between the more traditional poles of regional analysis carried out in archaeological survey and smaller-scale, site-oriented analyses through archaeological excavation. Greater methodological and theoretical sophistication at this scale of analysis is needed, given the amount of data collected from large-scale horizontal excavations and the intensive surface collections that are now more commonly implemented in archaeological research.

Gardens of Prehistory presents archaeological evidence for near-residential agriculture across a range of tropical and subtropical environments in the Americas (see fig. P-1 in Preface). The studies examine a variety of environmental contexts (from northern New Mexico to Central America) where prehispanic societies developed on the economic foundations of tropical cultigens and where our understanding of early agriculture is now developing or undergoing significant revision.

In the southwestern United States the record of both ancient and modern Puebloan agriculture has revealed a wide range of cultivation techniques practiced within and between settlements (Bradfield 1971; Cordell 1984; Fish and Fish 1984; Ford 1981; Forde 1931; Hack 1942; Masse 1979, 1981; Maxwell and Anschuetz, this volume; Woosley 1980). The existence of an equally diverse range of techniques employed within residential areas is now being established in the adjoining portions of northern Mexico (Doolittle 1980, this volume). To the south, in the highlands of central Mexico a variety of intensive agricultural strategies including intensive infield cultivation, terracing (Evans, this volume), and wetland field raising (*chinampa* agriculture) augmented shifting agriculture and floodplain irrigation as common components of highland agricultural production and settlement (Palerm 1955; Sanders, Parsons, and Santley 1979; Wolf 1959).

In lowland Mesoamerica an earlier perspective of low technological diversity and productivity among prehispanic agricultural systems has been altered greatly, especially in the Maya region (Harrison and Turner 1978; McAnany, this volume; Siemans and Puleston 1972; Wilken 1971; Turner and Miksicek 1984). It is now reasonable to hypothesize that similar early systems of intensive cultivation extended to the Gulf Coast lowlands as well (Santley, this volume; Killion, this volume; Santley, Ortiz, and Pool 1987; Siemans et al. 1988). Throughout the Gulf Coast and eastern lowlands of Mesoamerica the near residential component of prehispanic agriculture clearly would have represented a critical element of intensive and sustained subsistence production.

In Central America, and elsewhere, evidence for intensive cultivation practices close to habitation areas has been preserved below volcanic ash falls (Sheets 1983; Zier 1983, this volume), indicating the possibility of a more complex mix of agricultural strategies along the southern periphery of Mesoamerica's high culture area. In fact, evidence of such dryland field raising, now known from throughout the Americas, suggests a generic technique of intensive cultivation generally ignored in reconstructions of New World agriculture (however, see Fowler 1969 and Riley 1987). Prehistoric dryland "ridge and furrow" features began to be documented in both surface and excavated contexts in the last century (Avebury 1869; Hubbard 1878; Schoolcraft 1860) and have been identified subsequently in locations as diverse as central El Salvador and the Great Lakes region of North America (Fowler 1969; Riley 1987; Sheets 1983; Zier, this volume). This technique of cultivation, where earth is mounded in long rows and crops can be positioned either in the lower furrows or on top of the linear mound, seems to have served a variety of purposes including cold air drainage, excess moisture drainage, and possibly irrigation. Constructing these features, in many cases a regular part of the cultivation cycle, represents an investment in labor and an arguably intensive level of cultivation best conducted in proximity to residential areas. Ridge and furrow field systems seem to correspond well with evidence for population growth and complex social development from the area of Mississippian florescence in the midwestern United States to areas where Classic period societies became established along the southern Maya periphery in central El Salvador.

In South America, in addition to the existence of complex systems of irrigation and terracing long recognized in the Andes, recent research has documented the presence of prehistoric agricultural systems based on the transformation of marginal wetlands through field raising in both the highland (Denevan, Mathewson, and Knapp 1987; Erickson 1984; Kolata 1986) and the lowland contexts (Broadbent 1987; Denevan 1982; Farrington 1985b). The possibility of highly productive recessional cultivation systems for the floodplains of the Amazon (Meggers

1971; Roosevelt 1980) has likewise increased our understanding of the spatial mix of settlement and agriculture in a region generally assumed to be of lower agricultural potential. Ironically, accumulated data have generated a rise in expectations for the productive potential of prehistoric agricultural systems throughout the Americas just when contemporary awareness has been heightened by the imminent destruction of the environments within which these systems once thrived.

The theoretical emphasis of the present volume is to strengthen our understanding of the archaeological record as a consequence of the organization of agricultural production in and around settlements. Lithic and ceramic data are examined as diagnostic indicators of the organization of agricultural activities carried out in proximity to the residence (McAnany, chap. 8). Soil phosphate distributions and the location of relict agricultural features are likewise utilized to help identify the many uses of settled and cultivated space (Ball and Kelsay, chap. 10; Zier, chap. 9). Surface ceramic distributions are examined as indicators of the organization of agricultural field labor (Santley, chap. 7) and the use of urban space for agricultural and ceramic production (Killion, chap. 6). Maxwell and Anschuetz (chap. 3) and Doolittle (chap. 4) examine agricultural facilities or field features within near-residential and near-settlement or infield contexts. Evans (chap. 5) examines specialized crop production, settlement agriculture, and dietary requirements of late prehistoric populations in central Mexico.

In chapter 11, the concluding section of the volume, separate commentaries by Turner and Sanders set out their most recent thoughts on the subject of prehistoric settlement agriculture in light of more than thirty years of research in the field. The authors examine the role of settlement agriculture within the overall pattern of prehispanic agricultural production, its relationship to the spatial distribution of population, land tenure, and intensification, as well as the terminology, often confusing, that archaeologists have used to describe ancient agriculture. They go on to discuss and critique individual volume contributions from their own perspectives on the subject, outlining the extant strengths and weaknesses in the study of settlement agriculture and suggesting areas of theoretical and methodological concern for future research.

In sum, the articles collected in *Gardens of Prehistory: The Archaeology of Settlement Agriculture in Greater Mesoamerica* provide a new look at a critical element of food production in prehispanic America. The archaeological record is treated as a diagnostic indicator of the material consequence of some of the most redundant sorts of activities engaged in by farmers at varying distances from the residence, particularly those areas in proximity to the residence so important to the energetic parameters of prehistoric New World agricultural production. The outcome is a set of studies that seeks to close the gap between data and theory, provide new insight on the nature of ancient agriculture in the New World, and present some provocative methods and approaches that should help improve its documentation.

NOTES

1. By *extensive* I refer to systems of shifting agriculture generally practiced in areas of low population density where sufficient stands of climax and secondary forest resources are available for long fallow agriculture. This system usually, but certainly not always, implies a somewhat impermanent system of settlement as well.

2. The term *garden* refers to a polycultural mix of cultigens and useful economic species grown on small plots where the cultivator focuses on individual plants and their microhabitats by small inputs of labor on a continuous basis.

3. *Agriculture* is taken to imply a monocultural or near-monocultural mix of staple cultigens grown on large, flat, or landscaped parcels where the cultivator invests large amounts of labor in a staggered fashion during peak labor episodes of the cultivation cycle.

CHAPTER 2   WILLIAM T. SANDERS AND THOMAS W. KILLION

Factors Affecting Settlement Agriculture in the Ethnographic and Historic Record of Mesoamerica

A MAJOR FOCUS of this book is the house-lot or kitchen garden, a small plot of land located near the residence that has played a significant role in the agricultural systems and economies of a great variety of cultures in time and space. Gardens such as these, generally devoted to household production, are characterized by considerable variety with respect to plot structure, the crops grown, and their economic uses. Furthermore, as the discussion below seeks to elucidate, the function of household gardens must be seen in the context of their relationship to other agricultural fields, which are generally found at varying distances from the house lot. A number of writers, including those in this volume, have suggested that house-lot gardens in fact may have played a vital role in the very origins of agricultural economies. Hence, the house-lot garden is an important phenomenon critical to understanding both the *organization* and *evolution* of ancient agricultural systems and their dependent settlements.

In the sections to follow we focus primarily on the organizational aspects of agricultural systems, both economic and spatial, and make some general comments on factors affecting the form and function of what Killion (this volume) has called *settlement agriculture* or the cultivation of land located within and directly adjacent to settlement boundaries. These factors include its nutritional significance, relationship to the rest of the agricultural sector, and overall role in the economy of subsistence agriculturalists. We follow this discussion with some examples of historic and ethnographic settlement and agriculture from highland and lowland regions of Mexico and Guatemala. These examples provide a perspective on some of the potential variability present in the archaeological record of settlement agriculture in prehistoric Mesoamerica and the adjacent regions covered in this volume.

FACTORS INFLUENCING THE FORM AND FUNCTION OF SETTLEMENT AGRICULTURE

It is important to stress the fact that all subsistence farmers (i.e., farmers that produce most of the food they and their family consume) always cultivate a great variety of crops. Much polemic has entered into the literature because of the way in which this basic characteristic of many preindustrial agricultural systems has been applied to the archaeological record. Some of the difficulty relates to differences of scale and complexity among ethnographic models and the archaeological contexts to which they are applied. For example, the population densities achieved in tropical and subtropical America (i.e., much of Classic period Mesoamerica) and the cultivation practices probably implemented by many prehistoric farmers have few exact contemporary or ethnographic parallels. Even the most cursory examination of traditional American agricultural systems, however, underscores the adaptive importance of crop diversity regardless of technological complexity or demographic scale.

The reasons for crop diversity are clear: people are omnivorous animals, require a wide range of nutritional elements in their diet, and have a strong desire for variety, this last perhaps a physiological response to the first two traits. Nevertheless, inasmuch as the most important nutritional need, in terms of bulk, is caloric energy, these same subsistence cultivators derive the majority of their food intake from one or a few selected plants. These plants are referred to as staples and are derived from parts of plants that have unusually high yields of calories per weight, particularly seeds and roots.

Because such a high percentage of the diet derives from a few plants, field surfaces are generally dominated by these same few cultigens.

This requirement, to a lesser or greater extent, leads to an approximation of monocropping, a practice that puts high demands on soil nutrients. In the case of some staples, particularly those in which individual plants are relatively widely spaced, this condition may be ameliorated somewhat by intercropping. In Mesoamerica, for example, a common practice is that of intercropping beans and squash with maize. In cases where the staple is a small grain like wheat or barley and is most efficiently planted by broadcast sowing, intercropping is not feasible. Monoculture is also less of a problem with respect to the overall agricultural system through the cultivation of supplementary crops on field surfaces not generally devoted to staple production. The great variety of other crops that are required to fulfill the daily nutrient needs of the population but are not, or cannot, be intercropped either are cultivated in some sections of fields (generally dedicated to staples) or more commonly are planted in small parcels of land near residences. These often include herbaceous plants, grasses, and trees, the latter normally planted in separate spaces, in the form of groves, because of the damaging effect of their shade on smaller plants.

An additional factor complicating agriculture and land use that must be stressed is that most subsistence farmers do not live in socially or economically isolated communities. They are members of a larger society that is characterized by a socioeconomic heterogeneity, including differences in prestige, power, wealth, residential location, and economic specialization. These factors require that the cultivator must often contribute labor to other households, a situation that has to be somehow accommodated to his own household labor demands and, above all, articulated with the production of surpluses of agricultural products to pay rent and/or taxes or to sell in a local market context. These surpluses, in simpler versions of these economies, may parallel the range and ratios of crops produced for and consumed by the family, but they may also include considerable crop specialization, a feature that becomes more prevalent as the economy becomes more national or international. Today, in parts of southern Veracruz, Mexico, and Honduras, for example, small coffee groves—rather than the usual subsistence crop gardens—are found near the residences of subsistence cultivators.

The spatial relationship of the various kinds of agricultural fields to each other and to the residence is a complex matter, with a variety of factors influencing the relative nucleation or dispersal of settlement. From the point of view of work convenience and crop security, the ideal settlement arrangement for any subsistence cultivator would be one in which each residence is surrounded by all of its agricultural land, an arrangement that has been referred to elsewhere as *rancheria* settlement (Sanders 1957; see also Drennan 1988). Assuming that no other deranging factors are operating and that rough homogeneity exists in the characteristics of the land related to crop productivity, houses and fields would be evenly spaced over the landscape. Under conditions of a dispersed but dense population with more intensive land use, less differentiation would occur within the holding. Such a situation rarely pertains because deranging factors—such as variation in the productivity of the land, inequities of land tenure, and access to transport networks—always exist. This situation is particularly important when cultivators are heavily involved in marketing and hence are drawn into the economic and political arena of larger regions. In such cases settlement may still be dispersed but with considerable variation in population density and patterns of land use.

Finally, a variety of factors, often not directly related to agriculture itself, produce settlement patterns consisting of varying degrees of nucleation of food-producing households into discrete physical communities, and clear differentiation can be seen between a residential zone and the outlying agricultural fields. These factors, discussed below, affect the degree to which a farming population is nucleated into relatively large settlements and have important feedback relationships with how agriculture is conducted.

Socioeconomic. In many cultures land is owned by a corporate group, usually not of large size and usually consisting of a number of related families. Individual families often cultivate land separately, but even in these cases some farming tasks involve cooperative work that includes most or all of the adult members of the corporate group. Often such corporate groups form small physically isolated and nucleated settlements, which could be referred to as hamlets. In such cases small kitchen gardens are often present within the residential area and are owned and cultivated by individual households.

In some parts of the world, where private ownership and inequity of ownership are prevalent, much of the agricultural land may be in the hands of a few large landholders, a situation that is referred to as *latifundio* in Hispanic countries. In these situations the tenant farmers who cultivate the land often live in very large nucleated villages within or adjacent to the agricultural estate. These farmers have been referred to as urban villagers, and they are, in essence, a rural proletariat. An additional economic factor promoting communities of farmers in large nucleated settlements is a higher level of participation in the market economy, often involving not only specialized crop production and the sale of agricultural produce but also households in part-time craft specialization and services to other members of the community. Because of the greater number of nonagricultural or only partially agricultural community members and the multifaceted aspect of the household economy, it may be more convenient to live in villages and towns.

Technoenvironmental. In prehispanic Mesoamerica one of the most serious agricultural problems was that of maintenance of soil fertility. Maize is a highly productive staple but is a heavy feeder on soil nutrients, and this fact, combined with the poverty of the domestic animal sector of the economy, produced a serious problem for prehistoric cultivators, most importantly in the more humid regions where leaching produced an added problem in terms of soil fertility. These conditions acted as major stimulants to the development of house-lot gardens, because a garden placed near the residence would be more accessible in terms of the transport of exotic nutrients from the human and limited animal population occupying the residential lot.[1] Frequently, in areas of high population density in twentieth-century Mexico, maize is the major crop in such household gardens, and up to a third of the family's food requirements may be met from the production of this relatively small field. Angel Palerm referred to these as *calmilli*, or the "house maize-fields" (Palerm 1955). This factor, plus those discussed below, produced a settlement pattern with spatially discrete but low density villages.

Administrative Needs. Prior to the Spanish conquest, in some areas of Mesoamerica where population density was high, farmers resided in a completely dispersed settlement pattern or in hamlets. To facilitate conversion to Christianity and to conduct rituals more effectively, the church collected households in these areas and congregated them into relatively large nucleated settlements. The state also implemented the resettlement policy, in this case to facilitate taxation, and the resulting nucleated settlement pattern persists to this day over much of Mesoamerica. As already pointed out, these communities vary greatly in size and density. In many areas, while the village or residential area is clearly discernible and easily differentiated, the internal density might be quite low and intensively cropped small kitchen gardens or *calmilli* are found within the village area; outside the village are large fields, usually cropped using a fallowing regime. Eric R. Wolf (1959) refers to this pattern as infield-outfield agriculture and in some cases represents an adaptation of administrative needs to the problem of maintaining soil fertility mentioned above.

Military. An unsafe political environment may force cultivators to live in large, densely nucleated, often walled communities. In fortified settlements the constraint of wall construction may not allow space within the settlement for cultivation. If household gardens are present they would of necessity be outside the settlement area, possibly immediately outside its boundaries. Most of this infield zone (located directly adjacent to settlement fortifications) would, however, probably be reserved for staple crop production. This situation represents a case where the long-term farming decisions concerning fallowing and the maintenance of soil fertility take a back seat to the immediate needs of community defense, often at the expense of subsequent agricultural stability.

SELECTED EXAMPLES OF SETTLEMENT AGRICULTURE IN MESOAMERICA

To show how some of the above variables interact in particular cases, we will briefly present examples of twentieth-century Mexican (Sanders 1957; Charlton 1970) and Guatemalan (Stadelman 1940) peasant agriculture as illustrations.

Conquest period Mesoamerica exhibited a variety of settlement patterns, in part related to natural environmental factors that influenced the practice of agriculture, and in part due to other economic variables, particularly craft specialization and trade, and finally, political factors. In the Basin of Mexico an unusually high percentage (perhaps 25 to 30 percent) of the population of approximately 1 million people resided in large urban communities. These communities constituted densely nucleated settlements in which population ranged from approximately 3,000 up to the gigantic metropolis of Tenochtitlan-Tlateloco with its 160,000 to 200,000 people. A substantial percentage of people (perhaps an added 10 to 20 percent of the population of the Basin of Mexico) were subsistence farmers residing in smaller towns. The vast majority, however, of the purely rural settlements of the valley were very dispersed settlements located on sloping terrain. The density of these settlements was sufficiently low to allow the presence of large kitchen gardens as infields adjacent to the residence of the households, and these were undoubtedly used primarily for subsistence crops. Most of these fields, as they were located on sloping land, were terraced with the edges of the field protected by earth embankments, rows of maguey, or stone walls.

In many areas of Mesoamerica the elite population resided in central places the function of which was primarily political. The population concentrations in such central places were small and these centers had very limited economic functions. Primarily, they were places of consumption of goods from the surrounding countryside by the elite households and their direct dependents. Rural populations in these same polities, as a whole, tended to be quite dense but nucleated into hamlets, that is, nucleated settlements with a few score or hundreds of persons each. Between these extremes was a great range of settlement in terms of degree of urban development and of nucleation of the rural population.

Following the conquest two processes occurred that altered and, to a great extent, homogenized the patterns outlined above. The first was a process of population decline to perhaps only 10 to 15 percent of the conquest period numbers by 1615. The conquest period level of population density and size was not reached again until well into the nineteenth century as the aboriginal population gradually recovered from epidemic incursions of the sixteenth and early seventeenth centuries. The second was a process of nucleation of dispersed rural populations into large nucleated settlements. In areas where the population had declined strikingly or where the overall population was relatively small to begin with, almost all of the rural population was nucleated at the original central places. In other areas where populations survived in large numbers and where the original population was larger, people were congregated into rural but nucleated villages. This kind of pattern is referred to in the Mediterranean region as the urban village.

A considerable number of these Spanish-formulated settlements have survived into the twentieth century, particularly over the highlands of Mesoamerica and in northwestern Yucatan where the aboriginal population did survive to a great extent into recent times. Where the pattern did not survive was primarily in the tropical lowland regions where the population was almost wiped out by the great epidemics of the sixteenth century. The present-day population primarily represents a recent colonization from nearby highland areas. In these areas settlement tended to take on a completely dispersed pattern more in tune with the agricultural requirements. Sanders (1957) has referred to this kind of settlement in the twentieth century in the tropical lowlands as the *rancheria* pattern. To illustrate these points, we will use two examples from Sanders's own research, the Basin of Mexico in the central plateau and the coastal plain of Chontalpa in Tabasco. Finally, we discuss the patterns outlined by Stadelman (1940) for northeastern Guatemala. These patterns also can be compared to other ethnographically and historically known settlement patterns described by Hayden and Cannon (1984) for highland Chiapas, southern Veracruz (Killion 1990, this volume), and northern Veracruz (Kelly and Palerm 1952; Palerm 1955).

Basin of Mexico. In the Basin of Mexico today at least 90 percent of the peasant population lives in nucleated villages. Interestingly, considerable variation occurs in the internal density of such villages. Sanders (1970b) classified the villages of the Teotihuacan Valley sector of the Basin of Mexico in terms of this characteristic using the term *scattered villages* for communities that were nucleated, that is, there was a definite settlement area where the residences were located and then a larger area of outlying fields around them. In these communities the density of population within the village itself varied from 500 to 1,000 people per square kilometer with most of the villages clustering between 600 to 800 persons. In contrast, the term *compact village* was used for cases in which the densities were significantly higher, but it should also be noted that, in the Teotihuacan Valley at least, densities seem to cluster in two groups, one set of villages with a population density of between 1,200 and 2,000 people per square kilometer and another set between 4,000 and 5,000. In the *chinampa* (raised-field) area of the southern part of the Basin of Mexico many of the villages have densities as high as 8,000 to 10,000 people per square kilometer in the twentieth century.

While the explanation for this variety is complex, it seems to relate directly to the ratio of extended to nuclear families within the village (Charlton 1970). The higher density villages usually have a higher ratio of extended to nuclear families and these are also communities located in areas of high agricultural productivity and more severe land pressures. Villages with lower population densities tend to be those where nuclear families predominate as the basic household type and are located primarily in areas where the agricultural potential is lower, lands are poorer in quality, and pressures on land resources are somewhat less. What appears to be happening is that householders in the areas where pressures are high and land is valuable tend to hang on to their holdings after their sons and daughters reach marriageable age and maintain single households. On the basis of earlier studies (Sanders 1957, 1970a) it is argued that villages that are densely nucleated today tended to be nucleated rural settlements in 1519 and the villages that are now more diffuse represent Spanish resettlements of populations that were even more dispersed in prehispanic times. An additional variable that affects the nature of agriculture is the economic growth of Mexico City in terms of increased opportunity for the production of cash crops and part-time labor availability for villagers in the outlying settlements. Today many of the people living in such villages are no longer full-time farmers but practice a combination of rural and urban occupations.

In very densely nucleated villages the house lots are so small that they play a minimal role in the agricultural economy of the household. In such cases cultigens consist of a few fruit trees and a few flower beds or flowers planted in various types of containers along the front of the house. In a minority of cases in such villages where the house lots are somewhat larger, there may be as many as six or more varieties of such fruit trees, including both Spanish and native cultigens, and as many as twenty to thirty seedlings or fully grown trees in a single lot. Also typical of such house lots is the planting of *chayotes*, a vinelike plant that grows up the walls of houses and over the roofs, forming substantial and vigorous growths, and that has a high yield of fruit that is used as a cooked vegetable.

In more dispersed villages a considerable increase occurs in the variety of crops grown in the house lots and in their uses. Small kitchen gardens to produce a variety of native and European vegetables (primarily herbaceous plants) are present, and fruit trees are quite common, again including both native and European varieties, along with the ubiquitous flowers. Nopals are almost always found in such house lots and are perhaps the most common source of fruits and cooked green vegetables (the young leaves are edible).

Unlike the situation in northeastern Puebla described by Palerm (1952) and in highland Guatemala described by Stadelman (1940), maize is very rarely grown in house lots in the Basin of Mexico. Where villages are more closely integrated into the cash economy of the city, the nature of these household gardens shifts considerably from the very heterogeneous ones just described to gardens completely covered with a single monocrop, and in almost all cases this is a nopal orchard. With reference to the economy of households in the Basin of Mexico, the vast majority of the food consumed (at least 90 percent) is produced from the outlying fields, and because of the dense population and intensive use of the land, these "outfields" are rarely more than two or three kilometers from the residence.

Chontalpa, Tabasco. The second area we will use as a case example is the Chontalpa region of Tabasco, which contrasts very sharply with the situation in the Basin of Mexico. These observations date to the year 1953 (Sanders 1957), which is important to note because the construction of highways in the area has revolutionized the lives of the people since that time. The Chontalpa is a tropical lowland region in which the native population was virtually wiped out by the sixteenth century and the area was not recolonized significantly again until the nineteenth and twentieth centuries, primarily by populations from the nearby escarpment and highlands of central Mexico. Located on Mexico's southern gulf coast, the Chontalpa is a region of high temperatures and substantial rainfall (over 2,000 millimeters per year) and is an extremely complex area with respect to soil capabilities for subsistence or commercial cropping. A major river, the Grijalva, and a network of tributaries flow through what is essentially a flat coastal plain. The variations in the soils of the area are the product primarily of variations in the effect of the river and its annual floodings on nearby plains, and these variations play an extremely important role in the process of recolonization and the manner in which the land is used today. Particularly important in the economy of the region are three cash crops (cacao, bananas, and sugarcane), and virtually all producers are involved with varying degrees of their production. The peasants of the area classify the soils into three categories, *tierra de primera, tierra de segunda*, and *tierra de tercera*, based on the local farmer's understanding of soil texture and fertility. *Tierra de primera* is usually applied to areas of floodplain where the rivers flood periodically and where the soils are a sandy loam in texture and have a high natural fertility because of the annual flooding. These soils are ideal for raising cash crops and for producing subsistence crops like maize, beans, and rice. *Tierra de tercera* consists of a more acid interfluvial soil often so low in nutrients that instead of tropical forest, the natural vegetation that one would expect in areas of high rainfall like the Chontalpa, the vegetation consists of savannas. Very often strips of such land will have, prior to colonization, tropical forests in one section grading to savannas in another. *Tierra de segunda* is an intermediate category in terms of its soil characteristics and fertility and often consists of more hilly, better drained topography.

Much of the Chontalpa region up until the agrarian reforms of the 1920s was in the form of very large holdings used primarily for cash cropping. These lands were confiscated during the *ejido* reforms and the original owners were left with a maximum of one hundred hectares of land, much of which was in the *tierra de primera* part of the zone. The balance was broken up into *ejidos*, in this case, as parcels assigned to individual *ejidatarios* rather than held in common. The largest holdings are almost entirely planted today in the three commercial crops of sugarcane, cacao, and bananas. The owners of such parcels tend to reside in the small towns of the area such as Cardenas, Comalcalco, and Huimanguillo. In contrast, the small landholders, who are primarily *ejidatarios*, reside in a completely dispersed settlement pattern with each house located on its *ejido* parcel. These parcels are quite large in size compared to *ejidos* in the highlands of central Mexico because of the relatively low population occupying this area when the agrarian reforms were carried out. In the better soil areas six hectares of land were assigned to each *ejidatario* and on the poorer lands as many as twenty-four hectares. Some of these holdings, in particular the larger ones, are equal in size to what we have classified as a large landholding. Our distinction between large and small landholdings does not refer necessarily to parcel size or even to tenure relationship but rather to the way in which the land is used. Some of the *ejido* owners who received lands in the better part of the region have converted virtually all of their holdings to cash crops and we would classify them as large landowners because they are very often living entirely off a cash economy. Most of the *ejido* holders, however, combine cash crops with subsistence crops as their economic system.

Typically around each house in the Chontalpa is a small area consisting of perhaps two or three hectares of land in which the entire plot is converted into one of the three cash crops or sometimes a combination of two of them. For example, cacao is a small tree that requires a great deal of shade in order to grow, and bananas may be planted as a shade crop. More commonly a larger tree called the mother of cacao (*madre de cacao*) is planted to produce this effect. Also next to the house along with this commercial orchard is the typical kitchen garden, which we have described for other parts of Mesoamerica and in which a variety of herbaceous plants for vegetable production are grown. Also very common in the Chontalpa is another type of orchard, one that is very heterogeneous in nature with as many as twenty different cultivated trees of both native and European origin. Hence a house lot in the Chontalpa consists of the house, the cleared space for work, perhaps a small vegetable garden, a couple hectares of commercial crops, in which the variety is reduced to one or two crops, and then an orchard with a great number of species of trees in which the food is consumed by the householders but also sold as well.

Outside this core area around the house lot (where any distinct boundary between the zone of settlement agriculture and the infield is often blurred) is the major part of the holding usually split up into quarters because land pressure has caused a reduction in the fallowing regime. Of this infield area one quarter would be cropped in any one year and lands would rarely be rested more than two or three years. This generalization is more appropriate for the better quality lands; in the poorer lands a rotation system might have to be less intensive. A variety of subsistence crops is grown in these short fallow infields, often as many as thirty cultigens including, of course, the ubiquitous maize and beans, and very often a section of the field will be planted entirely in rice. Added to this variety would be a great number of root crops and other herbaceous plants such as chile peppers, squashes, and tomatoes. In other words, this is a typical swidden field with a great variety of plants characteristic of this type of agriculture. Because of the dispersed type of settlement in the Chontalpa, none of these infields that are planted in subsistence crops is more than five hundred meters from the house.

Highland Guatemala. Stadelman's study of maize cultivation in northwestern Guatemala (1940) contrasts with both the Gulf Coast and central Mexican examples outlined above in terms of environment, settlement configuration, and land-use strategies. The observations date from the late 1930s and provide a variety of data on the planting and cultivation of maize among traditional Mayan farming communities living in the Cuchumatan highlands of the Department of Huehuetenango, most of which at the time of the study could be reached only by foot and pack animal. This last example of settlement and farming practices in highland Guatemala probably represents the greatest amount of continuity in cultural traditions and demographic levels since prehistoric times of the examples discussed here. Stadelman notes that a primary objective of the study was to provide a reasonable approximation of agricultural practices in the area during prehistoric times when the region apparently supported relatively dense human settlement as well.

The Cuchumatan highlands are a rugged mountain region with isolated village farming communities located at elevations of from 750 to 3,000 meters. Annual temperatures range from 65 to 80° F within this elevational gradient with approximately 1,000 millimeters of rain falling primarily between March and October. Most settlement or agriculture in this area is located on steep sloping land; however, a number of villages and towns reside on and farm rich valley bottomlands. In general, settlement is fairly dispersed; regional population density averages about 25 persons per square kilometer with local densities varying widely between 8 and 800 persons per square kilometer. The traditional Maya population lives in three basic community types including larger villages with small numbers of ladino townfolk, smaller Indian villages surrounded by dispersed hamlets, and areas of completely dispersed hamlet and farmstead settlement. Farmland is owned privately by individual household heads but Stadelman believes that this arrangement is an adjustment to earlier communal tenure organization and related to a more recent state-imposed land tax system.

Maize, grown primarily for subsistence purposes, is the principal agricultural product of the region and is often intercropped with beans, squashes, and chiles. Most households raise a small number of sheep and chickens, and a few cattle and horses are kept in most communities for tillage and transport. Household members eat about two pounds of maize a day as tortillas or gruel *(pozole* and *bebidas),* and small amounts of meat, eggs, potatoes, tomatoes, wheat, and coffee are also consumed. Farmers dedicate most of their time to maize production but also supplement their income with migrant coffee and sugarcane labor at lower elevations. Stadelman estimates that, in general, community maize production would be insufficient to meet local needs so that cash for the purchase of food is a necessary component of the economy.

Within a community, agriculture is practiced in three basic zones corresponding to house-lot fields, infields, and outfields, all radiating out from the settlement area whether it be of the dispersed village or isolated hamlet type. Outlying fields are referred to as *montaña* (forest parcels) or *huatal* (bush-regrowth parcels) and consist of shifting cultivation plots that are cut, burned, and planted for two to three years and then allowed to rest for up to five years. Hoe, machete, and ax are the primary agricultural tools of the outlying fields. *Montaña* (freshly cut virgin forest) is the most productive class of land cultivated in the region and together with the *huatal* forms the largest areal component of the traditional Maya agricultural system of the Cuchumatan Mountains.

Inside this area of shifting agriculture is an infield zone of fields and settlement agriculture. The infield zone consists of two areas of intensive cultivation known as *llano* (short fallow sod land) and *pajonal* (grassland). These are permanent, occasionally plowed, short fallow fields that are fertilized with the ashes and residue of previous crops, green manure from weeding during cultivation, and sheep dung. Animals are penned in small corrals that are regularly moved over the field surface during fallow periods lasting up to two years. Hamlets and isolated farmsteads are sometimes found scattered among the highly productive *llano* and *pajonal* parcels, which, being more closely located to settlements, are considered the most desirable class of land and are often also terraced. Most residences, however, are found inside the innermost portion of the infield zone where *rastrajo* (cornstalk land) cultivation is practiced.

Rastrajo parcels, located directly adjacent to or within settlements, represent some of the most highly productive land cultivated by Maya farmers in the region. These plots are cultivated year after year without rest and receive constant inputs of fertilizer in the form of crop residues, green manure, animal manure, and all categories of organic household waste. In most settlements maize fields spread out between dwellings, filling up all the spaces between residences except for a small clear area or patio directly adjacent to the living structure. The more land devoted to *rastrajo* cultivation, the more dispersed the settlement. While areally the smallest component of the agricultural system in most communities, *rastrajo* receives as much or more labor input as any other component of the overall agricultural strategy and represents the most staple-oriented form of settlement agriculture discussed thus far.[2]

CONCLUDING REMARKS

Settlement and agricultural systems observed historically and ethnographically in Mesoamerica reveal considerable variation in crops utilized, land-use practices, and the spatial organization of field and residence. Basic human dietary needs require the sustained production of one or more staples in all agricultural systems; in Mesoamerica maize and, to a lesser extent, root crops have provided the bulk of carbohydrate inputs to subsistence. Protein needs are generally achieved on the basis of the favorable amino-acid complementation of maize and beans that are grown and consumed together throughout all areas of

Mesoamerica. Smaller amounts of protein are also provided by domestic animals, the contribution of which was greatly reduced by the poverty of this sector of the food-producing economy in prehispanic times.

Staples, while potentially cultivable on all portions of the agricultural landscape, are generally restricted to areas where larger field surfaces can be maintained. Supplementary nutritional requirements are met by a host of additional crops, the regular consumption of which provides necessary inputs of vitamins, minerals, and other nutrients fundamental to good health. These crops, potentially cultivated throughout a community's agricultural territory, are often grown on smaller field surfaces in proximity to the residence within a zone of settlement agriculture.

House-lot gardens and intrasettlement cultivation plots (which together form the zone of settlement agriculture), infields (adjacent to the settlement's residential zone), and more distant outlying fields constitute the potential spatial components of most Mesoamerican agricultural systems. These components are organized, to a great extent, in terms of the environmental and demographic parameters that exist in particular regions. The distribution of highly productive and more marginal farmlands, problems associated with the maintenance of soil fertility, and the energetics of small-scale farming systems that run predominantly on human labor sources are all mediated by local population densities affecting patterns of settlement dispersal and nucleation. Political and economic factors, however, also play an important role, further transforming the final configuration of settlement and system of agriculture practiced in any given area.

The ethnographic and historic examples of Mesoamerican settlement and agriculture discussed in this chapter are largely a product of political and economic transformations brought about during the five hundred years since the Spanish conquest. Certain factors affecting settlement and agriculture during this time, however, also would have been important during the prehispanic era. Settlement dispersal and nucleation, as noted above, are particularly important processes associated with demographic and environmental factors. The degree of nucleation, however, is also a product of basic farming energetics. Farmers generally attempt to locate themselves with respect to the most highly productive portions of the agricultural landscape so as to reduce the distance between residence and field and minimize daily energy expenditures. Domestic animals are ready sources of manure and, as in Stadelman's highland Guatemalan example, can improve fertility and reduce labor expenditures on agricultural plots located outside the zone of settlement agriculture. Drennan (1988) has recently observed that these factors acted to favor residential dispersal in densely populated landscapes throughout prehispanic Mesoamerica. Prehistoric Mesoamerican farmers were generally forced to maintain a substantial portion of the agricultural landscape in fallow in order to allow the soil to recuperate for future cultivation, and population dispersal was a natural solution to this problem. Under more crowded conditions the chemical improvement of nearby agricultural land could offset the problem of decreased forest reserves and shortened fallows. Lacking any large domestic animals, farmers had to rely on green manure and household wastes for these chemical improvements, and because fields located close to the residence were those most easily worked and fertilized, the zone of settlement agriculture and nearby infield holdings were of primary importance to prehistoric communities.

Even in cases where farmers had access to naturally fertile alluvium or artificially improved lands (irrigated parcels, raised fields, or terraces), the zone of settlement agriculture in most cases would have represented an area of primary importance to cultivation. Large-scale settlement nucleation in the archaeological record, therefore, probably represents more complex organization of the agricultural and nonagricultural portions of the economy and probably represents the exception rather than the rule for the urban societies of ancient Mesoamerica.

NOTES

1. *Exotic nutrients* refers to nutrients brought into the field from the outside and can include a variety of materials: animal wastes, plant wastes, chemical fertilizers, soil in solution from floodwater irrigation, nutrients in solution from floodwater irrigation or permanent irrigation. In the house-lot garden such inputs could consist mostly of plant, animal, and human wastes derived from household activities.

2. One outcome of the great range of variation in population density in northwestern farmlands is that farmers in less densely settled areas rely greatly on long fallow cultivation *(montaña* or *huatal)* while those in areas of more dense settlement focus on the cultivation of *rastrajo-llano/pajonal* land. Some farmers in the most densely settled areas migrate seasonally to forestlands to cultivate long fallow plots, further complicating the pattern of land use.

PART I

Settlement and Agriculture in the Arid Lands of Greater Mesoamerica

CHAPTER 3   TIMOTHY D. MAXWELL AND KURT F. ANSCHUETZ

The Southwestern Ethnographic Record and Prehistoric Agricultural Diversity

INVESTIGATING PREHISTORIC AGRICULTURAL SYSTEMS has been traditionally a focal point of archaeological research in the northern Southwest. Over the past one hundred years, these efforts have succeeded in making two significant contributions. The first is the identification of a variety of agricultural features, including rock-bordered grid complexes, terraces, checkdams, floodplain field locations, reservoirs, and irrigation canals (Woosley 1980). The second contribution has been the recognition that some of these technologies were highly labor intensive and others were land extensive (Lightfoot and Plog 1984). The appreciation of this variability has led researchers to question why some populations were more dependent upon intensive agricultural practices than other contemporary groups. Although considerable gains have been made in documenting local agricultural technologies, researchers have had less success in determining the conditions that led to intensified agriculture or land use. The failure of southwestern archaeologists to develop more robust explanatory models for the observed patterns is due to several fundamental conceptual and analytical shortcomings. Together, these deficiencies obscure important sources of diversity found in the archaeological record (Cordell and Plog 1979).

Comprehensive studies of prehistoric agricultural practices as part of settlement-subsistence systems have largely been limited to the Hohokam culture area of south-central Arizona (Bohrer 1970; Doyel and Plog 1980; Haury 1937, 1976; Hodge 1893; Masse 1979; 1981; Nicholas 1981; Turney 1929; also see Fish and Fish 1984) and the Chaco Canyon region of northwestern New Mexico (Hayes 1981; Vivian 1970, 1974, 1984). The Hohokam are known for their elaborate and extensive canal irrigation networks, terraced and gridded fields, contour terraces, and waffle gardens. Chaco Canyon, in the Anasazi culture area, features highly sophisticated runoff irrigation technology that includes ditches, dams, and earth-bordered gardens. Significantly, the co-occurrence of these complex farming technologies with elaborate architecture and exotic material remains has become the standard against which we measure cultural complexity and dependency on agricultural production.

Except for the efforts by Glassow (1977, 1980, 1984) in the Taos and Cimarron districts of northeastern New Mexico, no comprehensive archaeological surveys exist for agricultural systems outside the Hohokam and Chaco areas. Coupled with uncertain measures of what represents social complexity, this lack prevents us from even beginning to comprehend the technological and organizational structure of most local settlement-subsistence systems in the region.

Southwestern archaeologists have commonly assumed that canal irrigation technology, whether using existing water sources or surface runoff, represents a more elaborate social organization (see Dozier 1970:131–133; Eggan 1950; Wittfogel and Goldfrank 1943). Similarly, the absence of complex agricultural technologies is often interpreted as evidence of less developed social organization. This conceptual approach places us in danger of "assuming that if a complex irrigation farming technology is manifested at one location in an area, then the same people who built it could also not have constructed a simple, unirrigated, undammed, unraised field only a few kilometers away" (Woosley 1980:318). Many of these assumptions about past organizational complexity based on technological development are unfounded. For example, single-family units and other informal corporate groups among the Pima and Papago used complex irrigation technology (Castetter and Bell 1942; Hackenberg 1962). Ford (1972) also questions the link between irrigation and centralized authority among the modern

Rio Grande pueblos. More complex social interaction among the Rio Grande pueblos is required only when scheduling conflicts arise over the use of a shared canal-delivery system. The application of potentially inaccurate assumptions about the organizational complexity of a past population is further underscored by the failure of most archaeologists to examine adequately the effects of the natural and cultural environments on agricultural technologies. As Woosley (1980:317) points out, this implicit one-to-one mapping of a single kind of farming to a geographical area and to the people who inhabited it prehistorically may blind us to the variation that existed in their technological and organizational systems.

One final problem remains to be addressed—that of the ethnographic data base. Archaeologists have long been attracted to the detailed ethnographic data base available for the indigenous populations of the northern Southwest (Kidder 1924). They have relied heavily upon ethnographic analogies for definitions of the function of prehistoric farming devices and the organization of agricultural systems. Cordell and Plog (1979:407) argue, however, that these materials have not been assessed for their representativeness of the past.

As elaborated below, we question the general applicability of the ethnographic record in helping us to understand the structure and function of prehistoric agricultural systems identified in some local areas of the northern Southwest. Additionally, we argue that variability in contemporary agricultural practices of southwestern pueblos may not equal the variability observed in the archaeological record.

We believe that study of prehistoric agricultural technologies and land use in the northern Southwest must address inherent diversity in terms of adaptation. Variability in agricultural systems is related to highly localized environmental conditions, change within the natural and cultural environments, and the ability of local populations to respond to these changes. In contrast to the static reconstructions based on ethnographic analogy, this perspective views prehistoric agricultural technologies and land-use strategies as dynamic components of the population's settlement-subsistence systems. The goal of this chapter, therefore, is twofold. The first is to examine the adequacy of the ethnographic data base for interpreting the functions and structural organization of prehistoric agricultural features in the lower Rio Chama

Valley. We will further ask to what extent these ethnographic data can be used in assessing the roles that these fields and features played in the local prehistoric settlement-subsistence system. We also will examine the structural similarities between ethnographically recorded gardens and morphologically similar plots found in the archaeological record of the Rio Chama Valley. Therefore, our second goal is to present investigations in the lower Rio Chama Valley within a perspective that views the observed variability as a dynamic adaptive response of the local population to short-term variation in precipitation, temperature, and floodplain inundation.

EVALUATION OF THE SOUTHWESTERN ETHNOGRAPHIC RECORD

The ethnographic record for indigenous agricultural groups in the northern Southwest is a good starting point for the study of past settlement-subsistence systems in the lower Rio Chama Valley. This data base is, within limits, useful in identifying some major technological and organizational components of contemporary systems. Some insight into the basic parameters of the natural and cultural environments selecting for (or against) agricultural intensification and dependency may also be realized. This exercise specifically should not be treated as a search for some direct historical analogy to explain archaeological remains; the ethnographic present is not necessarily an appropriate representation of its past. Cordell and Plog (1979:407), for example, discuss how several innovative studies of the social organization of prehistoric Pueblo societies were seriously misled by their dependency on inappropriate ethnographic analogs. Better ethnographic models that help to explain the human behaviors represented in the archaeological record may be found among indigenous populations residing outside the Southwest.

Many Puebloan ethnographic accounts may be viewed, in part, as idealized narratives of societies that have been disrupted for centuries by the presence of Athabascan hunter-gatherers and Hispano- or Anglo-European settlers. In other words, the ethnographic data tell us how things would have been done under normal conditions in the past. Even in studies of some contemporary aspect of a cultural system, such as agriculture at Hopi (Anschuetz 1976; Beaglehole 1937; Bradfield 1971; Forde 1931; Hack 1942; Page 1940; Stewart 1940; Whiting 1939; Whiting, Jones, and Nequatewa 1935), informant data may more strongly represent idealized cultural rules for what should be done rather than what is actually achieved. For example, Forde (1931:393) reports that households at Hopi strive to maintain a greater part of a year's crop in reserve at any given time because of the threat of crop failure. Yet, this is an ideal that is seldom fulfilled.

A second factor affecting the usefulness of the ethnographic record is the biases of the recorder (Cordell and Plog 1979). Simply, the interests of the recorder will determine the kinds of information collected in a study. For example, an ethnographer interested in Pueblo religion may take great care collecting information about ritual observances and practices associated with ensuring successful harvests. Still, the details of how fields are selected, prepared, planted, and tended may be much less specific.

Ethnographic data may be used initially to identify departures in the archaeological record from a generalized model of contemporary behavior. This pattern of variability can then be used productively to refine the focus of archaeological investigations to determine the meaning of this variability. If used uncritically, ethnographic models can make the past resemble the present (Wobst 1978).

A Classificatory Framework Based on Southwestern Ethnographic Data

Two major classes of agricultural holdings—gardens and fields—are defined by the proximity of their location to permanent or seasonally occupied residential sites (fig. 3-1), the variety of plant species grown within a plot, and the amount of labor expended per unit of land. Gardens are typically found next to residential sites, feature a variety of specialty plants, and require great amounts of tending, including pot watering or canal water irrigation. Investment in formal features is high. Fields, on the other hand, are often located farther from residential sites, are generally monocropped, and require less daily tending except during critical times of the plants' life cycles. Fields may or may

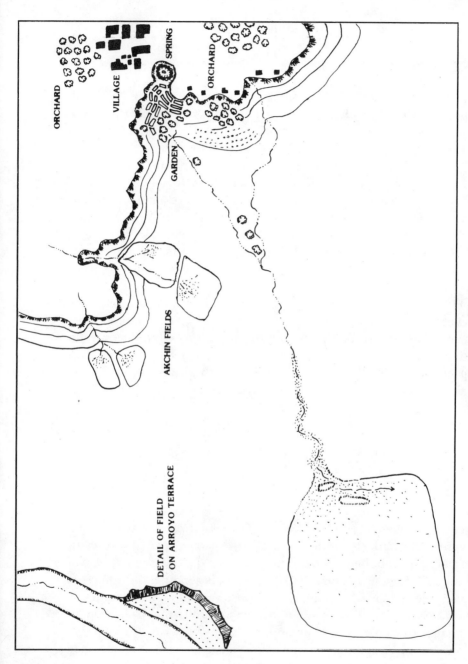

Figure 3-1. Schematic of the Relationship between a Hopi Village and Its Orchards, Gardens, and Fields (after Page 1940, not to scale)

Table 3.1 Hopi Garden Acreage

Groups	No. of Acres	No. of Operators
Tallahogan	4.5	9[a]
First Mesa	4.0	20
Bacabi	.3	20
Hotevilla	5.0	57
Lollama	.1	3
Burro	.1	4
Total	14.0	113

Source: Page 1940:68.
[a]Includes 4 Navajo.

not have specialized facilities that are designed to accommodate the needs of a specific plant or to manipulate distinctive aspects of the physical environment.

Gardens. Some indigenous southwestern farmers ensure minimal harvests, if only for a single family, with waffle gardens (Woosley 1980:321). These plots, which are small rectangular compartments enclosed by ridges of clay earth, have been comprehensively described for the Hopi (Forde 1931; Hack 1942; Page 1940; Stewart 1940) and the Zuni (Bohrer 1960; Cushing 1920; Stevenson 1904; Stewart 1940; Underhill 1946). Although gardens are noted at other pueblos, such as Taos (Smith 1967), Santa Clara (Hill 1982), and Cochiti (Lange 1959), these features have generally received little attention. In the Southwest, ethnographic gardens are commonly located next to the primary residential community or border residential room blocks. Smith (1967:8), however, notes that small gardens are also found next to "summer houses" at Taos. These residences, which are located outside the main community, appear to be the remnants of seasonally occupied farming settlements. Gardens are typically small but require large amounts of labor. Page (1940), for example, reports that in the 1930s only approximately fourteen acres of gardens were under cultivation by 113 operators at Hopi (table 3-1). Women are primarily responsible for the care of gardens, although older men sometimes assist in their maintenance and younger men help in their construction (Bohrer 1960; Page 1940).

Various plant species, many of them mesic species, are grown in

Table 3-2. Hopi Garden Crops

Major	Minor		Orchard
Chile	Beet	Lettuce	Almond
Onion	Cabbage	Pea	Apple
Sweet Corn	Carrot	Peanut	Apricot
Watermelon	Cauliflower	Potato	Cherry
	Coriander	Radish	Grapes
	Coxcomb	Safflower	Peach
	Cress	Tomato	Pear
	Cucumber	Turnip	

Source: Whiting 1939:12–13.

modern Hopi gardens (table 3-2). Although this produce constitutes only a small percentage of the general diet, it offers a nutritionally important variety of vitamins and minerals not otherwise provided by staple agricultural field crops (Anschuetz 1976:61–62). Page (1940:66), noting that many modern garden species have been introduced in recent centuries, suggests that green corn, lima beans, kidney beans, aztec beans, tepary beans, sunflower, squash, and cotton were likely the principal garden crops during prehistoric times.

One of the most important daily tasks in tending gardens during the height of the summer is the watering of the plants in the morning to keep them from wilting in the heat (Stevenson 1904:353). At Zuni, gardens are pot watered from walk-in wells excavated into the side of the adjacent riverbed (Bohrer 1960:182; Stevenson 1904). In contrast, the Hopi depend on small canal irrigation systems to water their gardens. Springs, which occur below the mesa rim near the villages, feed earthen and stone masonry reservoirs above the gardens. When the bung in a storage pool is removed, water trickles down a network of small ditches to the terraced garden complex (Page 1940:65).

Investment in formal garden features is high. Walk-in wells and spring-fed canal irrigation systems have already been identified as integral components of ethnographic Pueblo gardens (fig. 3-2). Other features include windbreaks, cobble borders and earthen dams, and waffle grids. Garden windbreaks may be jacal fences, such as those used at Zuni (Bohrer 1960:183), or strategically placed fruit trees such as

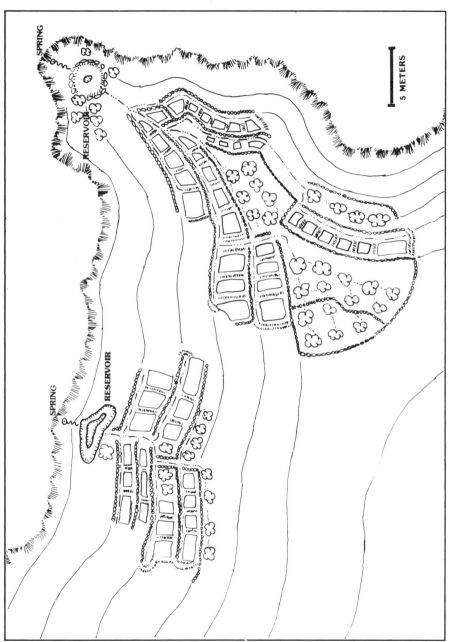

Figure 3-2. Schematic of a Spring-fed Hopi Garden Irrigation Complex (after Page 1940)

those noted at Hopi (Hack 1942:fig. 20; Page 1940). The cobble borders and earthen dams described at Hopi (Hack 1940:36–37) delineate not only the boundaries of a gardener's plot within the garden but also the planting rows and basins within the plots. The planting basins average seventy-five centimeters on a side. Waffle grids, which occur at Zuni (Bohrer 1960; Forde 1931; Stewart 1940), are highly formalized earth-bordered planting beds measuring forty-five to sixty centimeters on a side. The sides of the grids are ten centimeters high. The borders, whether composed of stone or packed earth, function both to hold irrigation water until it soaks into the ground and to protect the soil from erosion. Woosley (1980:321) notes that these features also serve as windbreaks that protect newly emerged seedlings from the effects of dry winds and conserve moisture by decreasing surface water evaporation during irrigation. These structures also would conserve available ground moisture by obstructing the movement of air across the planting beds between waterings, impeding the normal capillary action of soil moisture toward the surface.

Fields. Ethnographically documented fields include eight major types: dry-farm, seepage, water-table, slope-wash, *akchin,* floodplain, terrace, and irrigated (Bryan 1929; Hack 1942; Glassow 1980; Woodbury 1961). Two other classes of fields, linear-border (Glassow 1980) and gravel-mulched (Ellis 1970; Lang 1979, 1980, 1981; Luebben 1953; Tjaden 1979), are included in the present discussion, though they are known only through archaeological research. These fields are distinguished through edaphic and topographic criteria, and each type requires a different assumption about the degree of labor necessary for construction and maintenance. Edaphic variables relate directly to soil nutrients, composition, and texture, whereas topographic factors include slope, hydrology, and exposure. Various structural features, such as checkdams or grid borders, are found in association with certain field types. An examination of local environmental variables that affect agricultural production, including topography, hydrology, elevation, length of growing season, soil nutrients and composition, solar radiation, precipitation, temperature, frost, wind, hail, and predators (Ford 1972:3; Woosley 1980:318), is then employed to help explain the functional diversity among the documented features.

Fields are generally at some distance from residences although tremendous variability occurs. Some fields may be in proximity to residences while others may be a day's journey away. Comprehensive land-use analyses, however, have not yet been conducted in the northern Southwest, and we cannot currently discuss the differential distances between fields and settlements.

Dry-farming fields are planting locations that obtain their moisture directly from precipitation. Success in dry farming in any given locale is dependent upon the amount of precipitation the field receives, its slope, and the permeability and moisture retention characteristics of its soils (Glassow 1980:45; Hack 1942:32). At Hopi, which receives an average of only 250 millimeters of precipitation annually, dry farming is restricted to areas with a 30- to 100-centimeter-thick mantle of eolian sand that overlies less permeable subsoil. Significantly, the sand effectively acts as a dry mulch to conserve limited soil moisture (Hack 1942). In contrast, dry farming is possible at Mesa Verde because its greater yearly rainfall (450 millimeters) offsets the problems with less permeable clay soils (Glassow 1980).

Although the ethnographic record reports that several kinds of plants were grown in the dry-farmed dune fields (Hack 1942:32–33), Bradfield (1971:18) doubts that they could have played a significant long-term role in the Hopi agricultural strategy. At Hopi, the dune fields are susceptible to deflation and soil nutrient depletion. We suggest that, being prone to rapid nutrient exhaustion and long fallow periods, dry-farmed fields must be frequently shifted. Given this land-extensive characteristic, the fields are often some distance from settlements. Glassow (1980:45) uses the ethnographic evidence from Hopi to argue that dry farming contributed only a small percentage of the total agricultural yield prehistorically.

Seepage fields, as defined by Hack (1942:34), obtain a significant portion of their moisture through seeps or springs directly upslope of the plot. At Hopi, residential garden plots are often also located near seep or spring resources (fig. 3-2). Seepage fields, with their characteristically sloping topography, may also receive some quantities of surface runoff. Because they are not strictly dependent upon direct precipitation, seepage fields may be situated in a greater range of soils than their dry-farmed counterparts. These fields are restricted to a few areas scattered across the northern Southwest with favorable geologic, topographic, and climatic conditions. Therefore, Glassow (1980:46) contends that the amount of acreage watered by seeps and springs at any time would have been small. Although they have limited distribution, we believe that seepage fields may have assumed considerable local importance during short-term droughts that followed cycles of greater precipitation. In such circumstances, the water tables would still have been well charged. Glassow (1980:46) argues that the amount of labor required to build and maintain seepage fields would have been roughly equal to that for dry-farming plots.

Water-table fields are closely related to seepage plots because they also depend primarily upon groundwater for crop production. These fields derive their groundwater from high water tables found near streams, lakes, and rivers (Glassow 1980). Although this field type apparently did not exist among the historic Hopi following the severe episode of arroyo entrenchment of the late nineteenth century, various researchers believe that this field type would have been important during various times in the prehistoric sequence (Dean et al. 1985; Glassow 1980; Hack 1942:45–69; Schoenwetter and Dittert 1968). Glassow (1980:47), however, notes that under conditions of dense natural vegetation, flooding, and heavy soils, water-table fields may have been impractical given the level of prehistoric technology. On the other hand, Glassow believes that, under optimal conditions, water-table fields would not require any more labor investment than dry-farming or seepage fields.

Slope-wash fields were defined by Bryan (1929) as planting locations at the break of unmodified slopes along valley sides where runoff would naturally spread across the ground surface. Although Bryan did not identify slope-wash fields at Hopi, Hack (1942) includes them in his classificatory scheme. Glassow (1980:47–48) notes that because of the probability that many slopes would be too steep without artificial terracing, slope-wash fields were probably rare prehistorically. If they did exist, Glassow suggests that these plots likely required little labor investment.

Akchin fields compose most ethnographically documented flood-watered planting locations (Bradfield 1971; Bryan 1929; Glassow 1980; Hack 1942). *Akchin* fields are most visible on broad alluvial fans at the mouths of medium-sized canyons or arroyos where the gradient is low.

Water, no longer constrained by the drainage channel, can then spread across the ground surface. These fields are preferred at Hopi and Zuni (Hack 1942) because they deliver runoff from a much larger watershed and their stream flow characteristics may not require the construction of artificial spreaders. Because streams also deposit new sediments with low velocity flooding, fertility of *akchin* fields is maintained. At Hopi, *akchin* fields characteristically have a thin mantle of eolian sand, which may act as a dry mulch, overlying a clayey sand substratum that usually retains the available soil moisture (Bradfield 1971:17). Glassow (1980:48) notes that some *akchin* fields may also be in areas of low and undulating relief, such as the areas surrounding Chaco Canyon and Black Mesa. Depending on local topography and stream flow characteristics, *akchin* fields may require the construction of features such as spreaders and checkdams at the arroyo's mouth. These features slow the velocity of runoff and spread it evenly across the planting area (fig. 3-3). Stewart (1940:337) describes the use of elaborate herringbone patterns of dams at Zuni. During periods of flooding at Hopi, a farmer may visit his fields to dig shallow canals to direct water to certain parts of the plot (Hack 1942:28). The above features, however, are impermanent and need to be repaired or entirely rebuilt after heavy spring runoff or heavy flooding caused by violent thunderstorms. Therefore, *akchin* fields may be subdivided into several classes based on the amount of labor required for construction and maintenance. This would be a continuum ranging from simple systems in low velocity, low flow drainage settings, to complex fields in locales of fast, high stream flow (Glassow 1980:49).

Floodplain fields are situated along the floodplains of permanent or semipermanent streams, the low flood terraces of large arroyos, or within the bottoms of arroyos (Glassow 1980; Hack 1942). Features to spread water across the field or to slow its velocity are common in these settings. Still, the risk of crop damage is high, especially in the bottoms of arroyos. The amount of labor investment is again dependent upon local environmental conditions. Whereas Hack (1942:29) characterizes the stream floodplains as large *akchin*, Glassow (1980:50) argues that floodplain inundation is a qualitatively different phenomenon than an arroyo emptying onto its alluvial fan. Simply, Glassow contends that the flooding of perennial streams is generally much more severe than

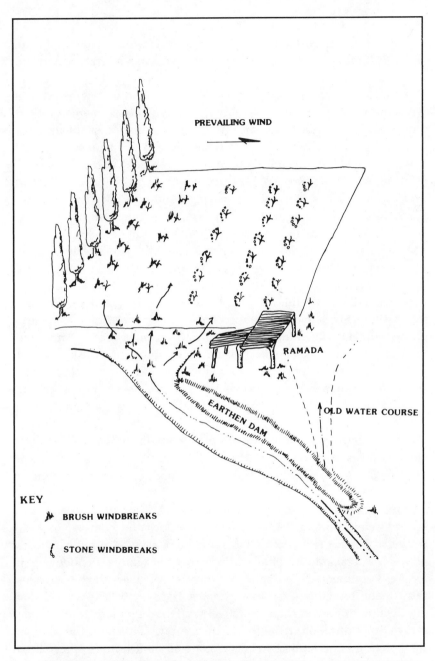

Figure 3-3. Hopi *Akchin* Field and Common Facilities (after Page 1940)

the flow of intermittent washes. Considering this line of argument, Glassow (1980:50) questions whether floodplain fields along most permanent streams in the Southwest were even practical in prehistoric times given the extant agricultural technologies.

At Hopi, *terrace fields* are defined as a series of rock checkdams placed at regular narrow intervals within the bottoms of small washes. These features control the velocity of floodwaters and protect arable soils from erosion (Hack 1942:30; Stewart and Donnelly 1943:134). Glassow (1980:51) contends that the concave cross-section for these drainage fields is a major diagnostic morphological characteristic. Given their formal plan and greater investment in durable features, these fields are distinguished from the *akchin* and floodwater fields discussed previously. Glassow (1980:51–52) believes that the construction of checkdams on these terrace fields would have required considerable labor investment. However, the level of technology employed would not have demanded any great expertise or elaborate social organization. In addition, once the fields were constructed, their seasonal and annual maintenance would have been minimal given their durable masonry construction. Therefore, Glassow (1980:52) reasons that in the long term terrace fields required the same labor expenditure as *akchin* fields equipped with less durable earth, stone, and brush dams and spreaders. Glassow (1980) unfortunately does not provide any data to support his arguments about labor investment.

Linear-border fields (Woodbury 1961:16) are composed of a series of parallel rock alignments that have been placed at intervals along the natural contours of a slope. The terrace walls may have multiple courses of stone masonry, or they may be composed of a single row of rock. Whereas terrace fields have a distinctly concave cross-section, linear-border plots have a convex profile (Glassow 1980:52). As such, linear-border fields are not likely to receive either the concentrated quantities or higher velocities of runoff characteristically captured by stream bed terraces. Although Glassow argues that linear-border fields have no exact historic Puebloan counterparts, Hack (1942:37) describes morphologically similar features within Hopi garden complexes.

Glassow (1980:53) argues that linear-border fields function to trap soils as well as surface runoff. These accumulated soils, in turn, would have been used as seed beds for agricultural crops. Archaeological evidence from Point of Pines (Woodbury 1961) and the Rio Puerco Valley northwest of Albuquerque (Moore 1981) suggests that linear-border fields were constructed in a wide range of topographic settings, including areas with steep (50 percent) slopes.

Glasgow (1980:53–54) further contends that these fields would have required substantial construction labor and that the effort needed to produce an acre of linear-border fields would have been roughly similar to that for terraces. He further equates the amount of labor necessary to produce a given unit of yield in these fields with terrace and *akchin* plots. Moore (1981), however, has shown that the labor investment in building linear-border fields may not be that great, and we believe that Glasgow may have overestimated the level of effort required.

Irrigated fields, which are common among the historic Eastern Pueblo (Dozier 1970), encompass a wide range of varieties. The essential characteristic of irrigated fields is the use of canals to carry water from areas of intermittent surface runoff, seeps or springs (Forde 1931; Hack 1942), or permanent rivers (Castetter and Bell 1942, among others). Archaeologically, such systems are identified at Chaco Canyon (Vivian 1970, 1974, 1984). The labor investment required to construct these systems may be considerable, especially if the feeder canals are supported by long main canals, reservoirs, and dams that require regular cleaning and repairs. As mentioned above, archaeologists often view canal-irrigated fields as evidence for complex social organization. Glasgow (1980:54) suggests that the energy costs would be significantly greater per unit of production than any of the other field types.

Gravel-mulched fields are reported to have been common prehistorically in the lower Rio Chama Valley (Ellis 1970; Lang 1979, 1980, 1981; Luebben 1953; Tjaden 1979) and near Picuris Pueblo near Taos (Gauthier, Prince, and Mathien 1978; Woodbury 1966). These fields, typically bordered with rock alignments, are characterized as planting beds with dense quantities of surface gravel. In some regions, cobbles are also used as a mulch. In the lower Rio Chama Valley, these fields are typically situated on former river terraces that overlook the floodplains of the permanent drainages.

Tjaden (1979:26–27) argues that the gravel mulch used to create these fields would have had two important functions. First, the rocky surface would have allowed the rapid infiltration of water from the surface. Second, the mulch would have acted to preserve available ground moisture by inhibiting evaporation from the soil (see Ellis 1970; Vivian 1974). More recently, Cordell (1984:206; Cordell, Earls, and Binford 1984:236) has argued that gravel mulches also serve to stabilize surface ground temperatures by reducing air movement near the ground and warming the first few centimeters of soil. Because the gravel absorbs solar radiation and dissipates heat more slowly than less rocky soils, Cordell suggests that the use of gravel mulches may have been a means to extend the growing season artificially. Although the gravel-mulched fields are not technologically complex, their initial construction would have presumably required the mobilization of a large labor force.

Discussion

In creating this classificatory scheme for gardens and fields from contemporary and archaeological sources, southwestern researchers have created a generalized model for the technological and organizational components of indigenous agricultural systems. Ethnographic data provide a useful basis to begin the formal examination of prehistoric agricultural systems, but the limitations of the ethnographic record are obvious.

The above classificatory scheme masks tremendous variability in the structure and use of fields and their facilities. To date, only rudimentary attempts have been made to understand the basic environmental parameters that lead to the selection of certain field types and feature assemblages. These limitations are particularly troublesome because the archaeological record of the Southwest encompasses even greater variability than the ethnographic record. Linear borders and gravel mulches do not have large-scale ethnographic counterparts. Although the seminal studies into the changing patterns of Hopi farming by Hack (1942) and Bradfield (1971) frequently serve as the foundation for archaeological interpretations of prehistoric agricultural features, these are static descriptions and not dynamic analyses of land use. Simply, we do not know how the ethnographic use of arable lands, fields, and field features changed through time and under what conditions. The lack of systematic and in-depth land-use analyses further limits our ability to assess the meaning of the diversity observed within the archaeological record through time and across space.

In dealing with these limitations of the ethnographic record, southwestern archaeologists have been prone to two fundamental errors. First, they have commonly relied upon an ethnographer's interpretation for the function of a certain feature without examining the degree of fit between the generalized model and the observation. Woosley (1980:329–32), for example, examined several series of checkdams near Taos and concluded that not all of the features were necessarily associated with agricultural activities. Rather, Woosley found that some of these features may actually trap water for domestic use. Before we can address questions of technology, structure, and diversity of prehistoric agricultural systems, archaeologists need to search the agronomic literature and devise systematic experiments to discover how fields, associated features, and their variants function under diverse conditions.

The second error involves the assumption that the predominant technology found in the traditional agricultural economy of an ethnographically documented population was also the dominant farming technique in the area during prehistoric times. Yet, very little of prehistoric Puebloan agriculture is visible in the ethnographic present. Pueblo farming systems today often consist of European-introduced crops, livestock, or cultivation technology. In addition, historical patterns of agriculture have been drastically impacted by European land-use policies, especially the establishment of reservation boundaries and the reliance on grazing livestock. In the northern Rio Grande, which includes the lower Rio Chama Valley, this error is most frequently expressed as the presumed use of canal irrigation technologies following A.D. 1200 (see Ellis 1970, 1975, 1979; Woosley 1980:325–26, 1986:159). This issue is important because there is a persistent belief that canal irrigation was a dominant component of the economy in the lower Rio Chama Valley during prehistoric times. We believe that this misconception has hindered the recognition and evaluation of the structural organization of the past settlement and subsistence systems of the area.

We do not doubt that canal irrigation was practiced in the northern

Rio Grande during prehistoric times; however, we question its role in local subsistence systems (after Anschuetz 1984:94–96; Cordell, Earls, and Binford 1984; Wozniak 1986). We suggest that the periodically volatile stream flow characteristics of the major permanent drainages (Ford 1972), dense vegetation and heavy alluvial soils (after Doolittle 1980), and the level of technological sophistication of Anasazi farmers inhibited the development of large-scale canal networks within floodplains. Instead, we believe that these populations had to depend more heavily on a variety of dry- and runoff-control farming technologies than has previously been appreciated.

The paucity of Spanish descriptions of irrigation systems prior to A.D. 1600 (de Espejo, in Hammond and Rey 1966:220; de Sosa, in Hammond and Rey 1966:282) suggests that prehistoric canal systems were not extensive. In addition, the archaeological documentation of canals (Bandelier 1890–1892; Ellis 1967, 1975; Ford 1981; Hewett 1953; Nelson 1914; Steen 1977) is restricted to lateral waterways of the major drainages; no prehistoric ditches have yet been found in the bottoms of the major drainages (Marshall 1982). These data further support our contention that canal systems were limited. This argument then compels us to search upland areas bordering the valleys to learn what other farming strategies were used by the prehistoric Anasazi populations in the northern Rio Grande.

In summary, we believe that the usefulness of the ethnographic data base in assessing the roles of fields and farming in local prehistoric settlement-subsistence systems is limited. It provides only a generalized model of agricultural systems for the early stages of archaeological research. Clearly, the ethnographic data cannot be used to explain the variability observed in the archaeological record, and the uncritical application of direct historical analogies may result in erroneous interpretations.

THE LOWER RIO CHAMA VALLEY

In 1986, the Museum of New Mexico, Office of Archaeological Studies (formerly the Laboratory of Anthropology Research Section), investigated four prehistoric agricultural sites in a small study area in the lower Rio Chama Valley (fig. 3-4). The sites, situated on the second and third terraces south of the Rio Chama, include the remains of extensive rock-bordered fields without gravel mulch, rock-bordered gravel-mulched and cobble-mulched fields, barrow pits, and possible reservoirs (Anschuetz, Maxwell, and Ware 1985; Anschuetz and Maxwell 1987). Some of these features, including groups of small cobble-bordered grids, morphologically resemble field and garden features described in the body of ethnographic literature for the northern Southwest. The fields occupy elevations between 1,805 and 1,850 meters, and the modern floodplain next to the fields averages 1,790 meters. All four sites contain Biscuit A and Biscuit B pottery, which date to A.D. 1350–1450 and 1425–1550, respectively (Warren 1979). The best dates for intensive use of these features is believed to be approximately A.D. 1350 to 1450 or 1500. In addition, a few sherds of Tewa Red, a pottery type that spans the sixteenth to the mid-nineteenth centuries, were found. This suggests that at least some fields may have been used intermittently during the early Historical period.

Anasazi Prehistory and Previous Research in the Lower Rio Chama Valley

Earlier sites have been recorded, but it appears that there was very little use of the lower Rio Chama by Anasazi populations prior to the thirteenth century. Regional chronologies cover two periods of interest in the lower Rio Chama: the Coalition period (A.D. 1200–1325) and the Classic period (A.D. 1325–1600). Survey data for the valley show that at A.D. 1200 no habitation sites existed. The only sites recorded for this period are eighteen scatters of lithic or ceramic artifacts. By A.D. 1250 four small permanent villages existed in the region and by A.D. 1350 ten pueblos were in existence (table 3-3). The establishment of new pueblos continued through the fifteenth century. Today, the village ruins contain from several hundred to several thousand rooms. By the early part of the Classic period some of the pueblos had grown to be some of the largest villages in the northern Rio Grande, if not the Southwest. Shortly after A.D. 1500, all of the region's villages were abandoned. Although a few field houses have been found, current data

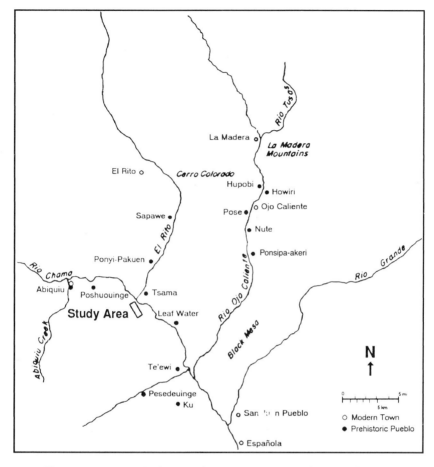

Figure 3-4. Major Prehistoric Sites in Lower Rio Chama Valley

show that no isolated residential units were built in the valley over the 300-year period from A.D. 1250 to A.D. 1550. The entire population was apparently living in the large villages.

The large Classic period settlements in the lower Rio Chama were the first sites to attract the interest of archaeologists concerned with the prehistory of the northern Rio Grande region. Bandelier (1890–1892) was the first European to describe the major communities, and in 1912,

Table 3-3. Prehistoric Pueblos in the Lower Rio Chama

Pueblo	Estimated Occupation Dates (A.D.)	Estimated Room Block Coverage (sq m)
Te'ewi	1250–1500	12,838
Tsama	1250–1500	20,255
Ponsipa-akeri	1250–1550	12,088
Leaf Water	1250–1375	1,174
Pose	1350–1550	10,629
Sapawe	1350–1550	26,917
Hupobi	1350–1550	11,154
Pesedeuinge	1350–1450	3,741
Nute	1350–1475	1,834
Ponyi-Pakuen	1350–1450	6,068
Ku	1375–1475	6,905
Abiquiu Canyon	1375–1500	244
Poshuouinge	1375–1475	11,350
Abiquiu	1375–1475	678
Howiri	1400–1525	20,860

the first excavations were undertaken by Jeançon (1912) at the ruins of Pesedeuinge. In 1919, Jeançon (1923) also excavated portions of Poshuouinge. Hewett (1906), Mera (1934), and Hibben (1937) drew maps of individual sites and plotted the regional distribution of large sites. Greenlee (n.d.) excavated portions of Tsama in the late 1920s and early 1930s, and tree-ring specimens were collected from some of the principal pueblos at this time (Hewett 1953). In the 1950s, Wendorf (1953) excavated portions of Te'ewi, and Luebben (1953) excavated a small part of Leaf Water. Florence H. Ellis conducted investigations at Sapawe and Tsama as part of the University of New Mexico Field School in the 1960s and early 1970s. Recent work at the Classic period pueblos consists of limited investigations at Ponsipa-akeri and its associated rock-bordered and gravel-mulched field complexes (Bugé n.d.a, n.d.b, 1981) and the excavation of a small portion of Howiri that lies within the U.S. 84 right-of-way (Fallon and Wening 1987).

Despite the seemingly intensive studies conducted in the area, the overt focus of much of the early research was to develop ceramic dating sequences, and little has been published with which to synthesize regional developments. Excavation data are also very sketchy, allowing little opportunity to assess the growth and internal organization of the pueblos. Wendorf and Reed (1955:150), citing a pattern of rapid changes in ceramic technology and decorative motifs during the early A.D. 1300s, argue that the abrupt growth in population indicates immigration into the valley. As further evidence they also refer to the appearance of certain items of material culture and various burial practices.

It has been further argued that a shift from an early linear village layout to the quadrangular "hollow square" pattern found in the later periods (Reed 1956) reflects fundamental changes in social organization (Peckham 1981:136–38). These changes resulted from increasing population density, the need to organize labor more efficiently, and possibly a need for defense. Peckham (1981:136–38) also contends that the growing regional population density required the development of previously marginal land for agriculture. Ellis (1975:20) argues that the observed aggregation of Anasazi populations into the large pueblos was necessary to provide the labor to meet the subsistence needs of that population. Others have argued that the placement of the communities on the top of mesas and benches suggests concern with defense (Hewett 1906; Hibben 1937; Mera 1934). Alternatively, it has been postulated that these locations would maximize available farmland in the floodplains (Dougherty 1980:12).

The complexes of rock-bordered and gravel- or cobble-mulched fields may have appeared in the valley soon after A.D. 1300, slightly later than the establishment of the first pueblos. A series of small gravel-mulched plots was found with a ten-room masonry unit (Fiero 1978), and Luebben (1953:15) found a series of rock-bordered plots next to the Coalition period component at Leaf Water pueblo.

These features, too, have been known for more than 100 years. Again, Bandelier (1890–1892) was the first European to identify the small agricultural plots near some of the larger pueblo ruins in the valley. He notes that the fields had frequently been misinterpreted by earlier explorers as the ruins of house foundations, an error repeated by later archaeologists such as Hewett (1906:36) and Greenlee (n.d.).

Greenlee (n.d.:17), however, was the first to note the association of circular depressions found along mesa edges with the misidentified "houses." He also observed that some cobble alignments appeared to lead to several of these depressions. Jeançon (1923:171) believed that the rectangular rock-bordered plots located about one kilometer east of the large pueblo of Poshuouinge were Pueblo shrines. After viewing aerial photographs of agricultural plots on top of Abiquiu Mesa, Hibben (1937:17) rejected Jeançon's functional interpretation and suggested that "a more obvious explanation would be that they are the remains of ancient maize terraces."

Settlements and Agricultural Features in the Lower Rio Chama

The major Classic period communities in the valley are located on old river terraces or basalt-capped mesas at elevations between 1,722 and 1,974 meters and, without exception, overlook the Rio Chama or its tributaries. Unfortunately, limited survey information prevents calculation of the amount of land surrounding each pueblo that was used for farming. Agricultural features in the valley include floodwater-irrigated fields, checkdams, rock-bordered fields without gravel mulch, rock-bordered fields with gravel mulch, field houses, barrow pits, and possible reservoirs.

Floodwater fields are known from the Ojo Caliente drainage and are found at the mouths of secondary drainages. Located in sandy soils, the fields resemble stone-outlined waffle gardens and contain small stone-bordered channels to distribute water from the arroyos to the plots (Bugé 1984:31). An area of more than 200 square meters near Sapawe is assumed to be a large floodwater field (Skinner 1965:18). Checkdams or small terraced fields have been found in several smaller arroyos draining into the Ojo Caliente (Bugé 1984:31), and Fallon and Wening (1987) report the occurrence of terraces (as defined by Woodbury 1961:11) and linear borders (Woodbury 1961:12) in the Ojo Caliente drainage.

Rock-bordered fields are well documented along the Ojo Caliente drainage. These fields are similar to linear-border fields but, in addition, have rock borders that run perpendicular to natural contours, serving to enclose the agricultural plot. These fields appear similar to the linear borders defined by Woodbury (1961:12) or the grids referred to by Plog and Garrett (1972:282). Glassow (1980:53) discusses the confusion between linear borders, grids, and terraces and concludes that the distinction lies in the way they collect water. He believes that linear borders tend to occur on "convex" landforms while terraces tend to occur on "concave" landforms. Bugé (1981:29) reports that these rock-bordered fields are the most common type of agricultural feature along the Ojo Caliente, and most are less than three meters on a side. He also reports finding larger, rectangular plots that are almost uniformly thirty meters on a side. These are found only where the slope is less than 5 percent.

In outline, *gravel-mulched or cobble-mulched fields* are similar to the rock-bordered fields but are covered with a layer of rock ranging from gravel to cobble size. Due to postconstruction erosion, these fields frequently appear as low platforms on the landscape. These field types lack a contemporary analog in the Southwest. They have been recorded by Bugé (1981) and, like those investigated by the New Mexico Museum's Office of Archaeological Studies, they are situated on old river terraces above the streams or on mesas overlooking drainages. Those next to the permanent drainages are generally no more than 50 meters above the floodplain. The features are not limited to the vicinity of major drainages; many are found on lateral tributaries where, today, no permanent water is found. The fields form an almost continuous band wherever remnant terraces overlook the drainages. As an example, a mix of rock-bordered and gravel-mulched fields is found extending in an unbroken line between the pueblos of Ponsipa-akeri and Pose, a distance of 9,700 meters. These fields cover an approximate area of 1.4 square kilometers. Unfortunately, uneven documentation of fields in other parts of the valley prevents any additional calculation of areal coverage.

It has been argued that the gravel-mulched fields occur more frequently on north-facing slopes to decrease moisture evaporation and protect crops that require less solar exposure (Fallon and Wening 1987:139). This argument cannot be supported with the current survey data. It is now known that extensive plots occur on southern slopes. It has also been determined that some fields investigated by the Office of Archaeological Studies are located where runoff cannot be collected.

These fields are situated on the highest points of both the second and third river terraces. In these positions, natural runoff cannot be collected or diverted even with the construction of water control devices.

Based only on the criterion of proximity to settlements, *gardens* are found within the lower Rio Chama Valley. Little morphological difference exists between these features and the larger fields. In general, an undifferentiated use of these rock-bordered and gravel-mulched agricultural features occurs with respect to elevation, exposure, runoff, or proximity to settlements. Luebben (1953) reports the presence of terraces, gravel-mulched gardens, and circular depressions at the Leaf Water site. The terraces and gardens are contiguous with the ruins, and they extend along the mesa edge away from the pueblo. At Ponsipa-akeri, Poshuouinge, Pose, and Tsama, corresponding features are found to abut or be in proximity to the room blocks. In contrast, the maximum distance between a major settlement and currently recorded fields is 4.6 kilometers.

Depressions ranging from 1.5 to 4 meters in diameter and up to 1.5 meters in depth are frequently found in association with these fields. These depressions are usually found on the downslope margins of the terraces but sometimes occur in the middle of the fields. Various investigators have suggested that these depressions may have served as barrow pits for the fill material used in the fields (Luebben 1953), as runoff reservoirs (Bugé 1984:32), or as locations for growing specialized crops (Bugé 1981:6). The depressions associated with the fields investigated during this study are generally positioned where little runoff could be collected. These depressions are usually located only 1 or 2 meters below the terrace borders and have catchment areas amounting to only a few square meters. Pollen data from the investigated depressions do not offer evidence for planting within the features. Based on drainage patterns and location, only one possible reservoir was found within the project study area.

Field houses are rare in the valley. Skinner (1965) has located twenty-four within 1.6 kilometers of Sapawe, but no others are recorded. The difficulty in discriminating small surface structures from a background of dense cobble deposits, particularly when the fields have internal, small rectangular partitions, may be part of the problem.

DISCUSSION

The wide-scale use of a diverse assemblage of agricultural features, one type of which is unique to the northern Rio Grande, cannot be explained simply by reference to the ethnographic record. Agricultural features comparable to the gravel-mulched field plots are lacking in the complement of agricultural techniques and features used by modern southwestern Indians. Although similarities exist, no features have the same properties as the gravel-mulched fields found in the lower Rio Chama.

Whether one considers the gravel-mulched plots in the lower Rio Chama to be gardens (based on proximity to residence) or fields, they share two characteristics: (1) they have a variety of internal features, and (2) they require a large labor investment in construction and maintenance. The diversity of internal features in the gravel-mulched fields may represent specialized functions. Excavation of plots (figs. 3-5 and 3-6) has revealed small, rock-bordered compartments very similar to Zuni waffle gardens. Unlike the rock borders around field perimeters, these borders are made up of cobbles aligned side-by-side rather than end-to-end, requiring more rock for construction. It is unknown whether specific plant species may have been grown within these cells, but such compartmentalized structures, along with other forms of internal partitioning within a single field, suggest the planting of diverse crops within an individual field. Other locations within investigated fields are sometimes characterized by the regular spacing of large flat boulders or elongated rock-bordered compartments. The function of these traits is unknown. It has been reported that large flat cobbles are sometimes used to grow flat-sided gourds for use as ceremonial rattles (Lange 1959), but palynological analysis of the fields (Clary 1987) does not offer supporting evidence. Therefore, the distinct morphological differences found within a single field (fig. 3-6) suggest the use of dissimilar growing strategies within an individual field location.

Construction of gravel-mulched plots also requires increased labor, and subsequent labor costs in crop maintenance are unlikely to be lowered. Gravel mulch acts as a barrier to evaporation loss but requires

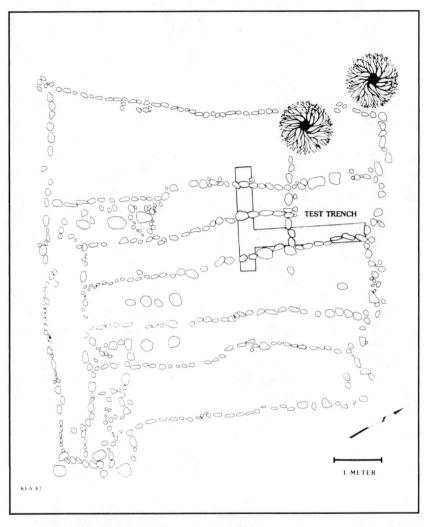

Figure 3-5. LA 48679; Study Unit 6, gravel-mulched plots showing location of excavation

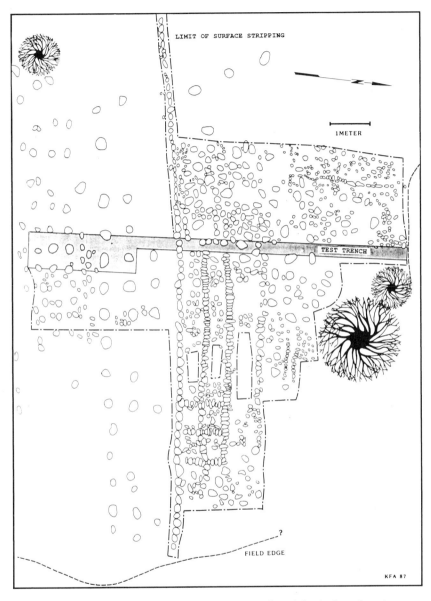

Figure 3-6. LA 48679; Study Unit 13, gravel-mulched plots showing limits of excavation

a great deal of upkeep. Soil water conservation studies by Lamb and Chapman (1943:577) show that a 65 percent stone cover results in significant reductions in evaporation rates and an increase in water absorption and helps to maintain a relatively high water-holding capacity. Corey and Kemper (1968:18) report that in tests using two different soil types, a clay loam and a fine sand, gravel mulch proved to be the most effective surface treatment for evaporation control and soil moisture maintenance. Fairbourn (1973:926) reports an eighteen millimeter per day constant rate evaporation in gravel-mulched fields compared with an eighty-eight millimeter per day evaporation rate on bare soil control plots. After four days, the water loss on the gravel field was still less than a single day's loss of moisture from the control plot. Fairbourn also reports that the gravel-mulched fields stored 60 percent of total winter precipitation (October–May), whereas the bare soil control plots stored only 40 percent of total winter precipitation. In assessing the effectiveness of gravel mulches in conserving soil moisture, Fairbourn (1973:925) concludes: "Large pores of the gravel permit rapid infiltration of water to the soil but retard evaporation. Water moves back to the atmosphere almost entirely by vapor phase across the gravel pores, thus the gravel tends to act as a one-way water valve for the soil."

However, agronomic research has shown that, to be effective, the gravel mulch must be renewed annually. To conserve soil moisture, the average pore space of the gravel mulch must be significantly larger than the underlying soil matrix. Corey and Kemper (1968:14) recommend: "The mulch should not contain a substantial fraction of grain-sizes that are smaller than the larger grains of the underlying soil. . . . In other words, nearly all the pores of the mulch should be larger than the largest pores of the soil below." Fairbourn (1973) shows that when wind and other natural processes cause gravel mulches to become mixed with soil particles, a significant drop in water conservation efficiency may occur. He cites lower growth rates for soybeans in unregenerated gravel mulches and suggests that annual regeneration of the mulch is necessary to maintain grain sorting and to maximize crop yields (1973:928). Rather than rejuvenate old fields, it may have been easier for prehistoric farmers to construct new ones. Therefore, an expansive distribution of field complexes along a river terrace may partially result from sequential construction, not from initially extensive land use.

The pollen data further suggest short-term use of the gravel-mulched fields. A high frequency of Cheno-am pollens from subsurface samples implies past soil disturbance resulting from agricultural activity (Clary 1987:13). Additionally, the low frequency of maize and Cheno-am pollen in other samples suggests that use of the fields was too brief for the pollen of cultivars and disturbance indicator species to leave a record (Clary 1987:16).

Significantly, it may not have been a long time before farmers were forced into less favorable growing locations. Three experimental growing plots planted by the authors and left untended, except for one weekly watering in a single plot, produced stunted plants of little nutritional value. Admittedly, close tending of the plots was not conducted, but the results suggest that plants grown in these plots still require more moisture than that derived from rainfall. Particularly in the hotter portion of the summer months, some form of pot watering or vegetative mulching may have been performed by prehistoric farmers during periods of low moisture availability. The plots are consistently found near drainages, even if they contain only intermittent streams. It is possible that the plots located next to streams with sporadic flow are of later construction. As plots in more favorable locations next to permanent water sources deteriorated, inhabitants would have been forced to construct plots in less advantageous situations, giving the appearance today of a past, extensive land-use strategy.

Another function of the gravel mulch may relate to temperature and plant growth. Agronomic field experiments suggest that the higher temperatures associated with gravel mulch may accelerate seed germination and plant development. Fairbourn (1973:927) reports that a 2–3° C increase in soil temperature under gravel-mulched fields resulted in significant differences in plant growth patterns. In contrast to cornstalk-mulched plots, the gravel-mulch treatment hastened germination and emergence of young corn by two to three days. He also notes that "plant growth rates were higher and mature plants were larger for all crops on gravel mulches when compared with other treatments. Corn tasseled 4–7 days earlier on gravel-mulched fields compared to all other treatments" (1973:927). Fairbourn concludes that conservation of soil water and increased soil temperature were the primary factors responsible for higher yields in gravel-mulched fields.

In the lower Rio Chama, the growing season can be short. The length of the frost-free season averages between 140 and 160 days, but wide variability occurs in the dates of first and last killing frosts (Reynolds 1956). Late frosts are not uncommon, and contemporary farmers in the valley report losing their late maturing crops to early frosts about every third year (Bugé 1984:32). The use of gravel mulch might provide a solution to both of these problems. The heat stored by the gravel might guard against early frost damage when crops, particularly maize, are most sensitive to temperature variation. Stored heat may also effectively lengthen the growing season by providing protection against late frosts.

The degree that lower Rio Chama Anasazi farmers were dependent upon the gravel-mulched fields cannot be fully assessed at this time given the lack of comprehensive survey data in the region. Although diversity within these fields is starting to be documented, the full range of variation in the region is yet to be charted. The use of floodplain fields has been documented (Bugé 1984:31; Skinner 1965:18), but these appear to be limited to the Ojo Caliente and El Rito tributaries. More than 200 years of historic farming have made the recognition of prehistoric fields difficult in the Rio Chama floodplain. Still, prehistoric farming in the Rio Chama floodplain was probably an uncertain endeavor. Hispanic documents show that early colonists were concerned with flood control and channel shifting as early as 1823 (Swadesh 1974:54). In addition, a flood destroyed the United States Geological Survey (USGS) monitoring station at Abiquiu in 1895 (Follansbee and Dean 1915:413). Significantly, all of the USGS reports for the late nineteenth century mention channel shifts throughout the lower valley (Follansbee and Dean 1915:408–15). The persistence of flood-related problems resulted in the construction of the El Vado flood-control dam in 1935. This evidence suggests that there may have been considerable difficulties in using the Rio Chama floodplain for agriculture.

Another difficulty facing prehistoric floodplain farmers may have been the dense stands of vegetation in the drainages. Thick stands of cottonwood and riparian vegetation would have required clearing before using the floodplain for agriculture. Although this vegetation could have been removed through burning, its disappearance would not have solved the problem of unpredictable flooding.

CONCLUSIONS

The diversity of agricultural features in the lower Rio Chama represents more localized variability in Anasazi agricultural production systems and strategies than is generally discussed. As Bugé (1984:34) postulates, this diversity may be related to risk minimization: risk is decreased by placing fields in diverse zones and water catchment areas. Should one production zone fail, another would serve as a buffer against total crop loss. Yet, the notion that this diversity is solely a result of risk minimization may be too simplistic.

It is unclear how farmers organize their subsistence behavior when they have access to different types of land and when they possess a broad suite of growing strategies. The observed diversity in the planting locations used prehistorically may also be the result of differential planting strategies over time. That is, all observed fields may not have been used simultaneously. An assortment of field types may be closely associated with localized environmental situations and may not be the proximate result of buffering tactics. Different environmental situations may have been used annually, leading to a pattern that is the consequence of accretional growth and not concurrent use. Further, planting strategies might have been conditioned by anticipated changes in the environment. For example, by monitoring the winter snow pack in the surrounding mountains, which are easily viewed from high points in the valley, inhabitants might have been able to forecast the volume of spring runoff. Should spring flooding have been anticipated, residents may have chosen to put more effort into farming at elevations higher than the floodplain. These changes in planting strategy can have important consequences for creating the archaeological record, and this is where most ethnographic studies fail. The dynamics of the system and how each component might receive differential emphasis from year to year is not well understood. It is necessary to determine these conditioning factors before assuming that observed diversity is the result of a single process.

Further, no clear morphological contrast exists between growing features that might be considered house lots, gardens, infields, or outfields. Features found immediately adjacent to residential lots closely resemble those found up to 4.6 kilometers away, indicating that in this case morphological considerations may not be useful for distinguishing the range of cultivation systems in the valley. That is, given that the features in the lower Rio Chama can be placed along a morphological continuum, it is difficult to use morphological criteria to determine whether a horticultural or agricultural emphasis may have existed in a specific location. Current palynological and phytolith data are also equivocal in this regard.

Given the existing data, there appear to be significant differences in the land-use patterns in the lower Rio Chama Valley when comparisons are made to patterns documented for Classic and Postclassic Mesoamerica. Data from the northern Southwest do not currently allow the ready distinction of classes of cultivation in a fashion similar to that used by Mesoamerican researchers. Most of the research conducted in the Southwest currently lags far behind that conducted in Mesoamerica, and more systematic, comprehensive land-use studies are clearly needed.

ACKNOWLEDGMENTS

Those who offered encouragement and help during this project include William L. Taylor and Steven A. Koczan of the New Mexico State Highway and Transportation Department, Andrew Darling, Robin A. Gould, Louann Haecker, Daisy Levine, Melody McLaughlin, Kevin McNamara, Ann Mehaffy, Sara Ann Noble, and David A. Phillips, Jr. We wish to thank them all.

CHAPTER 4   WILLIAM E. DOOLITTLE

House-Lot Gardens in the Gran Chichimeca
ETHNOGRAPHIC CAUSE FOR
ARCHAEOLOGICAL CONCERN

A RATHER SMALL but distinctive group of botanists (e.g., Anderson 1954) and geographers (e.g., Kimber 1966) has long been interested in house-lot gardens. For the most part, their studies have been limited to tropical areas characterized by a high diversity of plants and to modern times (e.g., Covich and Nickerson 1966). As a result of this regional and temporal focus, the existence and importance of house-lot gardens in arid lands have not been fully appreciated, especially by archaeologists. No one has yet to conduct a detailed investigation of present-day gardens in arid lands, and few archaeologists appear to have given their prehistoric counterparts much consideration. This condition is especially surprising for North America because house-lot gardens were probably used in the Southwest by 2000 B.C. (Yarnell 1965; Woodbury and Zubrow 1979:44). Indeed, they were used quite early in the humid eastern woodlands of what is now the United States (Fowler 1971:125–26; Ford 1981:8, 10), and it has long been recognized that parts of the Southwest probably received some cultivars, especially *Cucurbita moschata, C. pepo,* and some flint corns (Carter 1945; Smith 1987), from the east rather than from the south as is commonly thought.

The lack of attention afforded house-lot gardens is also surprising because they are not uncommon in arid areas. Furthermore, they are probably quite ancient. Indeed, several writers (e.g., Sauer 1952; Anderson 1952:136–50; Hawkes 1969; Bye 1979; Nabhan 1985) have suggested that house-lot gardens probably played a key role in the origins of agriculture.

That researchers working in arid lands have not given more consideration to gardens is most unfortunate. Archaeologists especially need to pay attention to house-lot gardens in such areas because preservation is often quite good, and our knowledge of desert cultures is probably incomplete because of this shortcoming. Evidence from other regions, such as Africa (Simoons 1965:15) and South America (Niñez 1986:9), suggests that house-lot gardens constitute an important part of the economic and social activities of many people. A better understanding of gardens should, therefore, provide valuable insight into such things as the role of women, children, the aged, and others whose daily activities do not take them far from, and are intricately linked to, the home (e.g., Harris 1971:180; Wilhelm 1975:92; Alcorn 1984:369).

There have been, of course, many ethnographic studies of arid land dwellers that have included information on gardens and gardening. Although these studies have to be perused, they often suggest the kinds of things that archaeologists should consider in their investigations. In this chapter, plants that were probably cultivated in house-lot gardens prehistorically in the Gran Chichimeca are first identified from ethnographic studies conducted in northwestern Mexico. Horticultural techniques employed in their cultivation are then reviewed. The material features associated with these gardens and the spatial relationships between houses and gardens are discussed next. Finally, survey data from one part of the region are investigated in order to confirm the use of house-lot gardens prehistorically and to suggest how relicts of such prehistoric features might be identified during future archaeological investigations.

The ethnographic data used here, especially those on plants and horticultural practices, come mainly from the Pima Bajo (Pennington 1980) of eastern Sonora, Mexico (fig. 4-1). Supplementary data, particularly on garden features built by the northern Tepehuan and Tarahumar who live in the Sierra Madres (Pennington 1963, 1969), are also used. The archaeological data employed here come from that part of eastern Sonora immediately north of and adjacent to the area inhabited

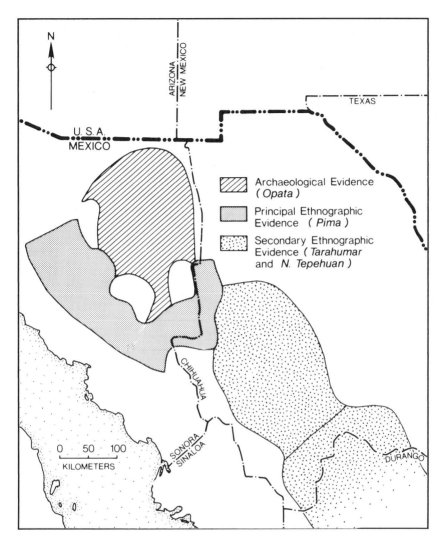

Figure 4-1. Map of the Gran Chichimeca Showing the Areas from Which Ethnographic Data on, and Archaeological Evidence of, House-Lot Gardens Are Discussed

by the Pima Bajo. This area was inhabited prehistorically by the ancestors of the now assimilated Opata, relatives of the Pima (Riley 1987). Most of the data were collected in the Valley of Sonora during 1977, 1978, 1981, and 1984 (Doolittle 1984c, 1988). These include community patterns and site features that have been reexamined for evidence of gardens.

GARDEN PLANTS AND THEIR USES

As with house-lot gardens everywhere, those of northwest Mexico comprise a number of different types of plants, of which various parts are used for numerous different reasons. Most of these were intentionally planted, but others are escapees and volunteers that are often protected because of their value (see Anderson 1954; Kimber 1973). Most gardens typically involve a mixture of fruit-bearing trees and shrubs, as well as plants grown for their leaves, roots, stems, and flowers (see Anderson 1952:137–41; Farrington and Urry 1985:147). Many plant parts are eaten, but others are used for medicinal, ceremonial, and other purposes (see Simoons 1965). Although garden plants could be discussed in any context, those found in the gardens of Indians in northwest Mexico are here discussed by their uses. Some Old World plants are evident in these gardens today. Because such plants would not have been grown in these gardens prehistorically, only those indigenous to the New World are covered here.

Pennington (1980:165–66) noted that gardens of the Pima Bajo differ from those of their Mestizo counterparts by having a fewer number of edible plants and a greater number of ornamentals. This observation suggests that prehistoric peoples in the area might have had a keen sense of the aesthetic, placing a high priority on visually attractive plants (Farrington and Urry 1985:146). Ornamentals found in Pima gardens include *Canna* sp., the climbing vine *Bougainvillea* sp., *Dahlia coccinea*, and *Zinnia elegans*. Not grown for their inherent beauty, but rather for the beauty that comes from their processing, are plants valued for their dyes. These include *Indigofera suffruticosa* and possibly even *Amaranthus cruentus*, which is also valued for its seeds and can be consumed as a green.

HOUSE-LOT GARDENS

In addition to the ornamentals, some plants are actually grown for other nondietary purposes. One tree commonly found in Pima house-lot gardens, *Piscidia mollis*, has bark that is used in the manufacture of a poison for the purpose of fish stupefaction. Two other plants, *Antigonon leptopus* and *Nicotiana tabacum*, are grown in order that their dried leaves can be smoked.

Many of the plants found in Pima house-lot gardens are, of course, grown specifically to be eaten. These vary from crops that constitute an essential part of the total diet to those that are probably used only for their taste. Of this latter type of plant, the most conspicuous are the chiles. *Capsicum annuum* tastes a variety of ways depending on when it is harvested. While still immature, these chiles are sometimes picked as the mild *chile verde*. Left to ripen, they become considerably hotter and are referred to as *chile colorado*. *Capsicum annuum* var. *minimum* has a small, pea-sized fruit that is extremely hot, especially when ripened and dried. These *chiltepins* are tasty but are so hot they can be consumed only in very small amounts. All the chiles are used as condiments, principally in bean and meat dishes.

Several Pima and a few Tarahumar garden plants are used as greens. In addition to *Amaranthus cruentus* (discussed earlier), *A. palmeri*, *Chenopodium album*, and *Lepidium viginicum* have leaves that are prepared by boiling and then consumed like spinach. *Amaranthus cruentus* is often found as an escapee, growing well in trash deposits; *A. palmeri* and *C. album*, on the other hand, grow in the wild and often invade household gardens. These plants are tolerated because they grow rapidly and profusely during the early part of the summer rainy season. This is the time of the year when the staple crops, especially maize, are in the field and stores from the previous year's harvest are running low. In effect, these *quelites* are used as an emergency food source during critical periods in years following poor harvests.

Pennington noted that an unidentified maguey, *Agave* sp., was cultivated in over half of all Pima gardens. After the leaves were cut, the bases were roasted in a pit before being consumed.

The greatest varieties of plants found in the house-lot gardens of the Pima Bajo involve those whose fruits and tubers are eaten. These include the trees *Pithecolobuim dulce*, *Achras zapota*, *Psiduim guajava*, *Lycopersicon esculentum*, *Cucurbita mixta*, and *C. moschata*. *Ipomea batatas*

73

is cultivated by the Pima while their Tarahumar neighbors cultivate *Solanum tuberosum* is their gardens. They also grow *Opuntia* sp. for its fruit.

Although not normally considered as house-lot garden crops, plants valued for their seeds or grain are not uncommon in northern Mexican gardens. In addition to *Amaranthus cruentus*, which is grown by the Pima Bajo, *A. retroflexus*, *A. palmeri*, and *Helianthus annus* are raised in house-lot gardens by the Tarahumar. Typically, the seeds of the *Amaranthus* are popped while those of the sunflower are dried, ground, and used in the making of small cakes.

There is no published ethnographic evidence of other seed plants being cultivated in either eastern Sonoran gardens or those in adjacent areas. *Zea mays*, however, has been found growing in many house-lot gardens there (fig. 4-2). A grain such as this could not produce a great deal of food and therefore might not seem like an important garden crop. This is especially so given that maize is the staple crop throughout the region and is typically grown in irrigated *milpas* and dry farmed in runoff-dependent *temporales* (Doolittle 1983, 1984a). The question then arises as to why anyone would grow a few plants in a garden when hundreds of hectares of the crop are planted nearby. The exact reason why present-day people grow maize in gardens is unknown. For prehistoric peoples, however, the answer might lie in the seeds themselves.

Fish, Fish, and Downum (1983:66–70) have recently proferred an intriguing interpretation of the function of the long-mysterious *cerros de trincheras* of Arizona and Sonora that might also provide a clue. According to them, the hillside terraces contain both house remains and fossil maize pollen. They argue that the terraces are not large enough for the ancient peoples to have grown a food crop. A small winter crop of maize, however, could have been grown on the hills because cold air drainage would have diminished the frost hazard. Such a crop would not have been large, but, unlike the summer crop grown in the lower elevations, it would be dependable year after year. Fish, Fish, and Downum argue that this small but reliable crop provided a constant source of seed regardless of what happened to the summer crop. House-lot gardens in Sonora are certainly different from the *trincheras* in a number of ways. They are, however, similar in that a small crop

Figure 4-2. Maize Growing in a House-Lot Garden a Short Distance North of the Town of Baviácora, Sonora

could be well tended and therefore depended on as a source of seed for the staple crop. The important factor here, of course, is the careful attention the plants must receive.

HORTICULTURAL TECHNIQUES

House-lot gardens are under constant and close attention of the people who own and use them, mainly because they spend a large part of every day in them. Indeed, anyone who has worked in or traveled through Mexico is well aware that much of the cooking, washing, child rearing, and even sleeping is done in residential lots—the gardens. Many of the plants found in house-lot gardens, therefore, get special and careful attention, and that attention is automatic by virtue of the plants' location in the center of human activity.

The most common form of deliberate attention afforded plants in northwest Mexican house-lot gardens is irrigation. For this purpose domestic wastewater is often used, as is water from nearby streams and springs. The Pima Bajo hand irrigate most of their ornamentals that are often grown in old or cracked *ollas*. They also water *Lycopersicon esculentum* and *Nicotiana tabacum*. No record exists of the Tarahumar irrigating this latter plant, but Tepehuan women have been known to hand irrigate it in their gardens. Although not documented, it might well be that contemporary and prehistoric peoples in the region hand irrigated *quelites*. Cheno-ams are ephemerals that do not sprout in dry years but sprout quickly and grow profusely at the onset of the summer rains during wet years. Supposing that hand irrigation was used on some crops and assuming that ancient inhabitants depended on *quelites* in the years following poor staple harvests, it seems likely that these people would have irrigated their *quelites* in order to ensure any food at all.

As has been noted, *Amaranthus cruentus* and all the cucurbits have been found growing in trash deposits. The connection between human refuse and gardens is a very old one (Anderson 1952:136–50). Also, numerous human activities, including the elimination of body wastes, typically occur in gardens (Kimber 1973:8). Dogs, turkeys, tree ducks, and a variety of wild animals are not, and in the past have not been,

HOUSE-LOT GARDENS

infrequent inhabitants of house lots and gardens in other parts of the New World (Kimber 1973:8; Linares 1976; Alcorn 1984:130–31). During the 1700s, one Jesuit missionary noted that parrots were often caged in Sonoran gardens (Pfefferkorn 1989:124). In sum, it would appear that gardens were well fertilized, by both human beings and other animals, even if explicit references to such an activity in the Gran Chichimeca are few.

Although irrigation and fertilization are most important activities in northern Mexico where water is scarce and the soils are low in organic matter, one cannot overlook the importance of assisting in the propagation and maintenance of the various plants. No documentary evidence exists of weeding being practiced by the Pima Bajo or the Tepehuan. That the Tarahumar weed their house-lot gardens, however, suggests that the practice is, and probably was, widespread.

Pennington (1980:165–66) noted that house-lot gardens of the Pima Bajo were a "jumble of vegetation." As is the case in other places where similar observations have been made (e.g., de Schlippe 1956), this observation is a clear reference to intercropping, the practice of growing a wide variety of plants together without any areal plan or organization. The advantages of intercropping are too numerous to list here. Suffice it to say, however, and for example, that trees act as trellises for some plants such as tomatoes that need support; squashes have broad leaves that shade the ground and thereby cool soil temperatures and reduce evaporation of scarce soil moisture; beans, being legumes, convert unusable atmospheric nitrogen into a form that can be absorbed through plant roots; and ornamental flowers attract numerous birds and insects that facilitate pollination of other garden plants.

The cultivation of such a wide variety of plants is not always an easy or straightforward task; some plants require special care. Most notable in this respect is planting. For example, before *Capsicum annuum* is planted, the Tarahumar soak the seeds in water for several days in order to soften the seed coat and facilitate germination. This soaking is usually done in *ollas*. Once the seeds are softened, the Tarahumar plant them in small beds for germination. After the sprouts are a few inches tall, they are transplanted into the main house-lot gardens. Also transplanted by the Tarahumar are *Lycopersicon* and *Nicotiana*. These plants, however, are not started in seed beds but rather are planted initially in baskets. Indeed, the baskets in which *Nicotiana* plants are started are either kept in the shade or are intentionally shaded.

The Tepehuan also transplant the same three plants that are so treated by the Tarahumar. Their practice is somewhat different, however, in that they do not plant initially in baskets. Instead they plant the seeds and protect the young sprouts in permanent seed beds. The Pima Bajo do not transplant as many different types of plants as do the Tarahumar and the Tepehuan. The only one documented as having been transplanted is *Nicotiana tabacum*. As with the Tarahumar, the Pima shade the germinating seeds. Unlike the Tarahumar, however, they do not plant in baskets but rather use seed beds much like those of the Tepehuan. The use of such features, especially ones that are shaded, involves the construction of permanent features.

GARDEN FEATURES OR HORTICULTURAL LANDFORMS

House-lot gardens in northwestern Mexico are typically small in areal extent. Pennington (1969:87, 1980:165), for example, states that Tepehuan gardens average approximately twelve square meters while those of the Pima Bajo typically measure about twenty square meters. This small size, however, might well be deceiving. Although little evidence exists to support this supposition, these gardens might be rotated in a manner not unlike that of swidden plots in a typical slash-and-burn agricultural system. For one thing, being both small and well demarcated, these gardens are certainly distinctive from their tropical counterparts that are often large and almost indistinguishable from nonutilized vegetation in the house lot (e.g., Wilhelm 1975:82; Alcorn 1984:115) and even the surrounding natural vegetation (e.g., Kimber 1973:8). Furthermore, the Tepehuan are known to move their corrals (used for penning large Old World animals) periodically and plant new gardens in the well-fertilized abandoned corral sites (Pennington 1969:72, fig. 6). The possibility of this practice having evolved out of relocating refuse middens of dump heaps (e.g., Anderson 1952:136–50) in prehistoric times cannot be ruled out.

HOUSE-LOT GARDENS

The practice of demarcating gardens, or at least having distinguishable features evident on the landscape, might well be a great advantage for archaeologists. Unlike gardens in tropical areas, those of arid northwestern Mexico typically have some type of rock enclosure. Indeed, a common practice involves the building of a low stone wall two to four courses high. Enclosures this low function primarily as deterrents to pedestrian traffic, and they probably keep very young children from damaging the plants. Low walls also assist in the containment of mulch materials that otherwise could blow or be washed away.

Although he does not record them in any of his books, Campbell W. Pennington (personal communication, 1987) has said that the Tepehuan, the Tarahumar, and the Pima Bajo also often have garden plots located in the bottoms of gullies immediately below their mesa- or hilltop houses. The downstream sides of these gardens are typically bounded by a low stone wall that actually traps soil and organic matter, especially household refuse and human excrement. Such gardens have also been seen as far north as the present-day city of Cananea, Sonora (fig. 4-3). Furthermore, traces of similar, albeit more elaborate, but now abandoned garden terraces occur behind and below houses in the town of Baviácora (fig. 4-4).

Similar in many respects to the low rock garden enclosures are the seed beds used for the germination of some plants. Such features are distinctive, however, in that they are usually quite narrow, rarely more than one meter wide (Pennington 1969:92, fig. 14). The purpose of making them so narrow is to facilitate the tending of young plants. The size of seed beds is not important as seeds can be planted quite close together if they are later to be transplanted. Accordingly, the construction of a narrow bed allows for the cultivator to reach several rows of seeds or sprouts without reaching very far.

Another factor that distinguishes seed beds from garden plots is the earth within the stone walls. Whereas the gardens themselves tend to have soil at the same elevation within the walls as on the outside, seed beds tend to be filled nearly to the top with earth. This fill is not, however, identical to the earth outside the bed. Indeed, it is usually very high in organic matter and has a greater moisture-retention capacity.

Figure 4-3. A House-Lot Garden in a Gully below a Mesa-Top Settlement near Cananea, Sonora. Note the

Figure 4-4. Abandoned Garden Terraces on a Mesa Slope below Houses in Baviácora, Sonora

SPATIAL RELATIONSHIPS

Although not an "agricultural landform," a modification of surface relief related to cultivation (Golomb and Eder 1964), there is one other important characteristic of house-lot gardens in northwestern Mexico—their locations relative to the houses themselves. My own observations have revealed that present-day gardens in the region tend to be located on the south sides of houses. Such locations might be intended to protect young plants from the northerly winds in the early spring. They might, however, have more to do with the availability of summer sunshine. By locating gardens on the north or shady side of houses, horticulturalists would be reducing the number of hours the gardens would be in the sun each day.

In addition to being located on the south sides of houses, house-lot gardens in northwestern Mexico tend to be located in places slightly lower in elevation than, or downslope from, their owners' houses. Such locations are, undoubtedly, a function of drainage and runoff. Rainfall in the region is, of course, low. During the late summer, however, heavy thunderstorms are common and can result in large amounts of runoff in brief periods of time. Locating gardens below houses helps prevent flooding of the dwellings and facilitates the washing away of household refuse and watering of the crops.

Identification of the plants, horticultural techniques, and especially the physical features associated with the house-lot gardens of indigenous people in northwestern Mexico today and in historic times helps in suggesting the types of things archaeologists should consider when conducting future investigations in the Gran Chichimeca. Although the region as a whole is often perceived as having been inhabited by barbaric nomads, there are many locales, such as the valleys of eastern Sonora, where sedentary agriculturalists did reside (Doolittle 1981, 1984b). These areas have not received a great deal of attention by archaeologists because of a general lack of interest, due in part to early surveys that found little evidence of permanently settled farmers (e.g., Bandelier 1890–1892), and, more recently, because the governmental policies of Mexico make archaeological exploration difficult. There have been some studies—most notably a survey conducted in the Valley of

Sonora (Doolittle 1988)—that provide evidence to suggest that household gardens were probably quite common in prehistoric times.

SURVEY EVIDENCE OF PREHISTORIC HOUSE-LOT GARDENS

In total, 162 prehistoric settlements or sites used for permanent habitation during prehispanic times have been found in the Valley of Sonora (Doolittle 1984c:19). Most of these sites are located on mesa tops overlooking the floodplains of the Rio Sonora and its major tributary arroyos. Evidence of houses-in-pits, dating from A.D. 1000 to 1550, and, especially, above-ground pueblo-like surface structures, dating from A.D. 1250 to 1550, is abundant (Doolittle 1988:36–38). No less than 1,289 prehistoric dwellings have been identified. House remains are not, of course, the only features on these sites; other identifiable items exist, such as roasting pits and defensive walls. There are also several features that had functions yet to be explained or that have been only partially explained. Although the possibility was not considered earlier, it might well be that many of these features were related to house-lot gardens.

Among the most conspicuous yet mystifying of these features are rings of rocks set into the ground endwise and adjacent to each other (fig. 4-5). A total of ten such features from six sites have been found in Sonora. Most are only about one meter in diameter, but some are nearly two meters across. These features have been thought to be the foundations of bulb-shaped thatch and plaster granaries like those found in caves farther east in the Sierra Madres (Sayles 1936:19). This interpretation, however, is at best speculative. No other evidence exists for such granaries in Sonora, and no one has assessed the foundations of those in the mountain caves. While the storage function remains to be verified, it is clear that these rings were not used as hearths. No evidence of charcoal has appeared in those that have been excavated, and none of the rocks shows signs of having been burned.

Given the problems with prior interpretations, it might be wise to consider that these rings were used in horticultural pursuits. One possible function might have been for the growing of ornamentals. Such features have not been reported in the ethnographic literature of north-

Figure 4-5. Rock Ring at Site Son K:4:20 OU. This ring could have served as a bed for growing ornamentals

ern Mexico. Somewhat similar rings of rocks used for the protection of ornamentals have been seen near houses farther south in the central Mexican village of Santa Caterina, Guanajuato. That the prehistoric rock rings in eastern Sonora were used for the growing of ornamentals tends to be supported by their locations; all of the known rings are proximal to house remains. It is also feasible that these rock rings could have been used as seed beds for germinating crops that eventually would be transplanted.

The possibility of seed beds had not been considered in previous archaeological studies in Sonora. However, some tantalizing evidence exists for their use. On at least one site, Son K:4:46 OU, two parallel rows of rocks were found, each approximately six meters long, only about one meter apart (fig. 4-6). These rock alignments were also found slightly downslope from the houses and too close to the edge of a steep escarpment to have been remnants of walls from two houses set very close together. The ends of these walls are currently not connected, but erosion along the top of the narrow mesa could certainly have destroyed any end walls. Given their location on the site itself, and their morphological similarity to currently used features, these alignments could quite feasibly be remnants of an ancient seed bed.

The evidence for seed beds is admittedly speculative. That for house-lot gardens themselves, however, appears to be more substantial. Seven sites in the Valley of Sonora have low walls of rocks a course or two high, often ten to fifteen meters in length (fig. 4-7). These walls are located parallel to the edges of the mesa tops and perpendicular to the gentle mesa-top slopes. Either by intentional fill or by natural depositional processes, these walls have collected sediment, resulting in a wide, flat surface suitable for horticulture. Presumably, these areas would have had fertile soil. With few exceptions, these walls are located downslope of houses where household refuse would be carried by runoff. Most are also located south of the houses.

These walls are certainly too low to have been used for defensive purposes. They could, of course, have been used for leveling the site surface for house construction. By definition, however, a terrace built for a house could also be built for a household garden. As Fish, Fish, and Downum (1983) have demonstrated, terraces suitable for houses can be cultivated.

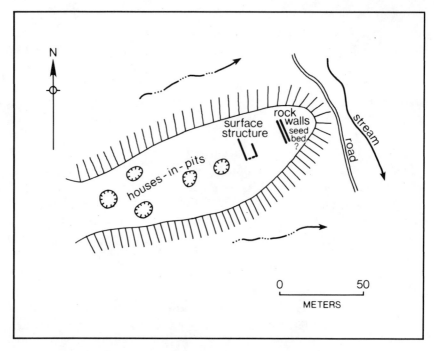

Figure 4-6. Two parallel Rock Alignments at Site Son K:4:46 OU. These features could be the remains of an ancient seed bed.

Of all the archaeological evidence collected in eastern Sonora, that which has the greatest similarity to house-lot gardens used in nearby areas today are low rock enclosures. Thirteen such features have been found on seven sites. These enclosures are typically two to four courses high, rectangular in areal configuration, and measure approximately four by five meters on each side (fig. 4-8). Given their morphological similarity to rock enclosures of Tepehuan gardens (Pennington 1969:88, fig. 13), it seems plausible that these features are remains of house-lot gardens. Not inconsequentially, all of these features are near other relict house foundations, and at four sites they are downslope of the confirmed houses (fig. 4-9).

As has been stated, there is a tendency for horticulturalists today, and there appears to have been a tendency for them in times past, to locate gardens downslope from houses. Given this finding, it might be

HOUSE-LOT GARDENS

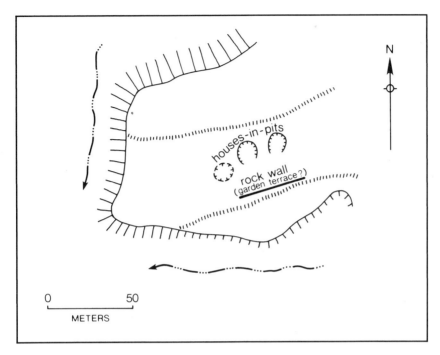

Figure 4-7. Rock Wall at Site Son K:4:118 OU. Note that the wall is parallel to the edge of the mesa top and downslope of three houses-in-pits.

inferred that the locations of the houses themselves might be evidence of house-lot gardens. A review of the *rancherías* or smaller permanent habitation sites in the Valley of Sonora revealed that this premise has some validity. For this portion of the study, settlements that had nine or more houses were not considered on the basis that larger sites would have relatively less space available for gardens than would smaller sites and that the locational characteristics of the houses on the mesa tops would be much more uniform for larger sites than for smaller ones.

Most of the 130 *rancherías* found so far in the valley have been damaged to some extent by either historic or present-day occupation, erosion, or recent road construction. Indeed only 38 sites appear to have evidence of all of their late prehistoric structures visible on the surface. An analysis of these sites revealed that most were located along the north side of the mesa tops overlooking arroyos (fig. 4-10).

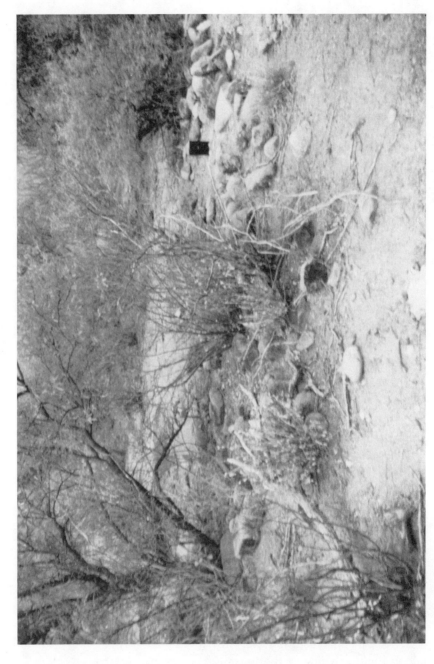

Figure 4-8. Rectangular Rock Enclosure at Site Son K:4:48 OU. This enclosure might be evidence of a

HOUSE-LOT GARDENS

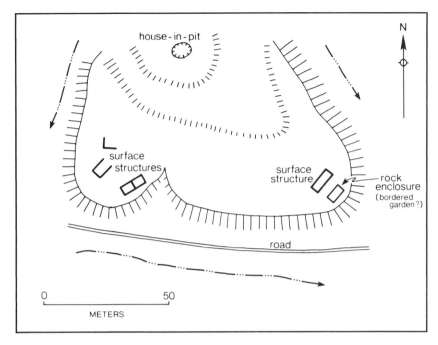

Figure 4-9. Rock Enclosure at Site Son K:4:144 OU. Note that the low rock enclosure is next to and downslope of the house foundation in the far southeastern corner of the site, at the edge of the mesa.

Given that the mesa tops where sites tend to be located are not flat, but slope slightly toward the river and toward the arroyos on the south sides of the mesas, 32 (or 84 percent) of the intact *rancherías* have most of their houses on the northern and upslope sides of the mesas. Clearly, there was a tendency not to build on the downslope portions of the mesas. It was in these areas that house-lot gardens might well have been located.

CONCLUSION

This study has demonstrated two things. First, house-lot gardens indeed are being used today in northern Mexico, and they are cul-

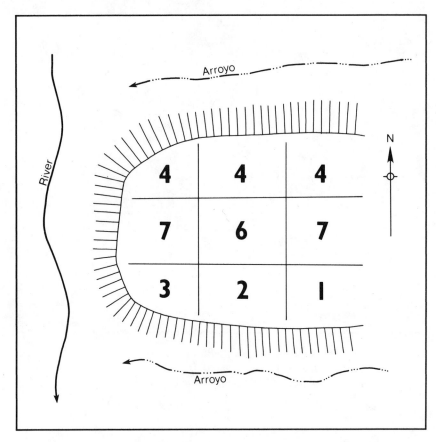

Figure 4-10. Composite Map Showing a Stylized Mesa Top Bounded on One Side by the River and on Two Sides by Tributary Arroyos. The numbers in the grids indicate the number of *rancherías* in the Valley of Sonora that had most of their houses in that particular location on their respective individual mesas.

turally important features that beg further and more in-depth study. Second, good evidence exists to indicate that similar gardens were used in the region prehistorically. They too require further investigation.

The systematic study of prehistoric house-lot gardens is actually in infant stages of development. Although botanists and geographers have long been interested in present-day gardens, most of their studies have been conducted in tropical areas where plant diversity is great.

For the study of prehistoric gardens, archaeologists working in arid lands where preservation is great now have the opportunity to make considerable advances in our understanding of horticulture and its cultural implications.

The previous pages have outlined some of the things to look for during archaeological surveys. Most notable in this respect are unusual rock features located in proximity to houses, especially on their downslope and south sides. Also not to be overlooked are open areas in these same locales. Although physical evidence of gardens might not always be immediately apparent, such vacant places might be indicative of ancient horticultural plots.

Future studies should look for such evidence. They should also go one step further and systematically assess suspicious features and areas. Specifically, the next generation of prehistoric garden studies should involve the analysis of sediments for chemical alterations due to, and biotic remains of, cultivation. For example, sediment analyses such as those developed by Robert C. Eidt (1984) can be used to distinguish cultivated from uncultivated areas and, hence, determine whether or not a specific place once had a garden. Similarly, identification of opal phytoliths found in sediments should prove helpful in identifying the types of plants grown in gardens (Rovner 1983). It is doubtful that pollen analysis will prove to be of any value in the study of household gardens. Pollen is simply too buoyant and blows too far too easily to be of any use. Phytoliths on the other hand form in all parts of the plant and are deposited in situ as the plant itself decays.

The data on prehistoric house-lot gardens in arid lands in general, and in northwest Mexico specifically, are admittedly tentative and the interpretations speculative. This fact does not, however, diminish their importance. Although more research is greatly needed, ethnographic evidence does suggest that gardens did exist there and that they played a significant role in the lives of the ancient inhabitants.

ACKNOWLEDGMENTS

I thank Campbell W. Pennington and Clarissa T. Kimber for their comments on earlier drafts of this chapter and Nancy Stanford McCarry for her assistance.

CHAPTER 5	SUSAN T. EVANS

The Productivity of Maguey Terrace Agriculture in Central Mexico During the Aztec Period

CARL O. SAUER once wrote that maguey permitted settled life in the arid central highlands of Mexico. Where no other water or food is available, maguey (agave, century plant; *Agave atrovirens* and others) thrives and sustains human life, and beyond this subsistence role the plant provides medicine, fiber, building material, fuel, and even fertilizer from its ashes. Maguey cultivators in the Basin of Mexico during the Late Postclassic period (or Late Horizon, ca. A.D. 1150 to 1521) pioneered the more agriculturally marginal parts of the environment (such as the sloping piedmont zone around the alluvial plain) with a land-use strategy involving terraced interplantings of maguey and grain over the hillsides of their villages. Probably established in the twelfth and thirteenth centuries, these villages covered the piedmont of the Teotihuacan Valley, Texcoco piedmont, and similar areas of the Basin of Mexico by the time of Spanish conquest. Archaeological and ethnographic evidence permits us to reconstruct maguey's contribution to the caloric needs of the maguey farmers and to apply this productivity value to an archaeological case, the Aztec period village of Cihuatecpan in the Teotihuacan Valley.[1] This reconstruction is based on modern maguey yields and ancient settlement patterns.

The resulting model of Cihuatecpan's caloric requirements and productivity demonstrates the village's adaptive success in using this mixed agricultural strategy. Maguey farming permitted the village to be calorically self-sustaining, and the sap offered necessary liquid refreshment in a semiarid zone while adding to the nutritional balance. The plant also provided the raw materials for a variety of crafts. Maguey was a crucial part of the economic base of this and many similar villages, and it secured Cihuatecpan's economic prosperity, evident in archaeological remains such as well-made, nicely finished houses. In spite of the marginality of this environment, the village could depend on its own land to support itself and also produce surplus goods.

HISTORICAL BACKGROUND

The large and complex structure of Aztec period society in Mexico rested on a base of peasant productivity. Feeding and fueling the huge capital, Tenochtitlan, and other large cities were the jobs of rural villagers who worked the land of the Basin of Mexico (fig. 5-1). Intensification practices and exploitation of locally available resources gave these peasant villagers a mixed economy, in many areas ensuring local prosperity while providing the larger society with valuable goods and services.

The flow of goods and services through the peasant sector and upward ramified and intensified as sociopolitical complexity increased in the last century before Spanish conquest. The fifteenth and early sixteenth centuries were a time of maturity for these villages, which had been established in the Late Toltec period.

Population growth was derived from intrinsic increase and from migration into the basin during this period (A.D. 950–1150; Sanders, Parsons, and Santley 1979:93, 137–49). The peasant villages founded then made use of thinly settled parts of the basin that were relatively marginal for grain cultivation when undeveloped but that had considerable productive potential, given application of intensification practices such as drainage agriculture (in the swampy southern lakes of the basin) and terracing on the sloping piedmont surrounding the basin's alluvial plain.

These peasant villages were the building blocks of the cultural system's foundation, the fundamental units of production for the basic

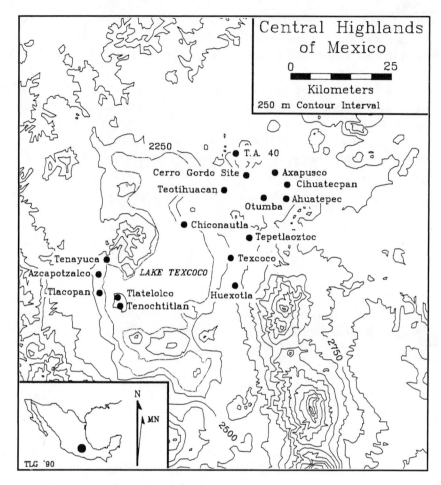

Figure 5-1. Central Highlands of Mexico

goods and labor of the society, and political control of the rural hinterland was an important prize in the ongoing struggles among the most powerful polities of the basin. It is important to note that prosperity of rural villages depended on the success of their strategies of land use, and these show an ingenious interplay of economic production and habitation.

In the terraced villages of the semiarid Teotihuacan Valley, for example, remains of house mounds distributed over the village's sloping arable land (fig. 5-2) indicate a probable pattern of small house-lot gardens (called *calmilli,* sing. *calmil*) surrounding the houses, similar to the dispersed house-lot land-use pattern known ethnographically from modern villages in the same kind of environment. If the ethnographic analogy can be extended, then the Cihuatecpanecos probably also had other, larger fields (in this report called *milpas*), which were also part of the village's hillslope land but were not adjacent to their houses.[2]

Cihuatecpan's pattern of houses on terraced lots reached its greatest extent in the early sixteenth century, perhaps 500 years after the village's establishment (according to ceramic chronology, see Evans and Abrams 1988, and obsidian hydration evidence, see Evans and Freter 1989). The village was probably founded by one of the many bands of migrants from north of the Basin of Mexico who had abandoned their villages when Toltec power declined.[3] These displaced village agriculturalists began to drift south to areas of greater security, to the Basin of Mexico and its adjacent zones to the east and west (Calnek 1982).

The pioneers found many areas of the basin already settled with some towns of several thousand people and small hamlets in the countryside. The density of settlement was light in comparison with that during the Late Aztec and Early Colonial periods, and potentially productive zones were not yet developed into habitable and arable land. Development and settlement of these more marginal lands proceeded apace, and the migrants succeeded in ingeniously using many such areas. The city of Tenochtitlan provides an excellent example of this phenomenon; established in an uninhabited swamp in the fourteenth century (or possibly before; Reyes Corts and Garcia-Barcena 1979), its population was 150,000 to 200,000 in 1520 (Calnek 1973:190). Other examples include the development of large-scale *chinampa* zones (intensive cultivation and settlement) in the marshy lakes of the southern basin (Armillas 1971; Parsons 1976) and the establishment of terraced villages described above (Parsons 1971; Sanders 1965; Sanders, Parsons, and Santley 1979).

The expanses of Aztec period terraced settlement along the Teotihuacan Valley's piedmont apparently began as a discrete set of small Late Toltec period settlements, according to the findings of the Teotihuacan Valley Project (Sanders 1965). Late Toltec sherds usually

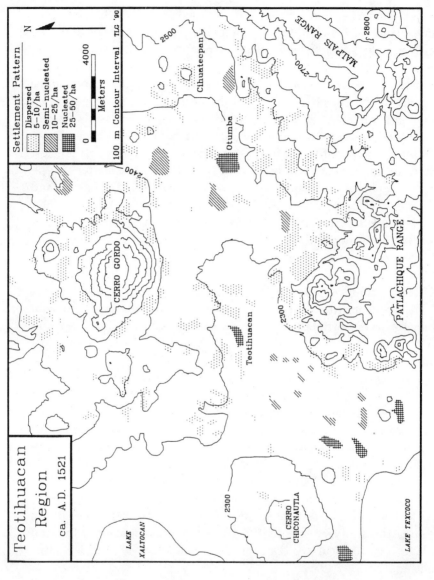

Figure 5-2. Teotihuacan Region Settlement Pattern (ca. A.D. 1521), showing range of persons per hectare

constitute a very small component of the ceramic collection from Aztec period villages, but they are found in the core areas of these dispersed villages and evidence the pioneering phase (see also, for example, the Cerro Gordo site; Evans 1985). The lack of Aztec period dispersed settlement over the alluvial plain might indicate that even in the Toltec period the alluvial plain was reserved for agriculture, with only a few nucleated settlements. The ethnohistorically known Late Aztec pattern of dividing this area into estates for nobles or for the support of public offices was probably established hundreds of years earlier.

Some time after A.D. 1000 the village of Cihuatecpan was established: small quantities of Late Toltec period wares (Mazapan and Atlatongo types; Evans 1986:314) were found in the surface and subsurface collections from many of the mounds on the south and southwestern slopes. Cihuatecpan was thereafter continuously occupied as a village until the beginning of the seventeenth century. Judging from ceramic material in direct association with architecture, the period of greatest population at Cihuatecpan was in the fifteenth and early sixteenth centuries, when it was a village of about 200 houses, encircling the lower slope of Cerro San Lucas, a small volcanic cone (fig. 5-3).

Excavations at Cihuatecpan revealed evidence of several specialized craft activities, in particular obsidian tool production and maguey processing. Several extensive dumps of gray-black obsidian debitage were found (one, measuring twenty-five by thirty-five meters, is described in Abrams 1988), probably from the Otumba (Estetes) source three kilometers away or from the barrancas nearer the village, where cobbles are readily available. Gray-black obsidian from this area is too highly textured to lend itself well to prismatic blade production, but it is good material for the uniface and biface knives and scrapers that were found at the site (Evans 1988:45) and no doubt were traded away as well.

Gray-black obsidian endscrapers were common and are thought to have been used for maguey sap production. Production of maguey beverages (fresh *aguamiel* and fermented *pulque*) involved collecting the sap from the plants (probably by sucking it into gourds) and then transporting and storing the liquid in large jars. Sherds from large storage jars are common in the village, typically with a flaring neck and direct rim (see Evans 1988 and Evans and Abrams 1988 for utility vessel

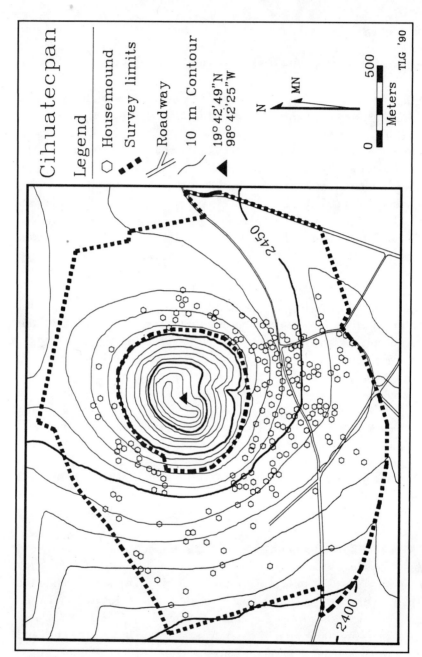

Figure 5-3. Cerro San Lucas and the Village of Cihuatecpan (hexagons = house mounds; dotted line = limits of survey)

frequencies). Body sherds from jars have not yet been analyzed to determine whether they were used to store maguey beverages.

Basalt scrapers shaped like modern hoe blades were common at Cihuatecpan, found on the surface and in excavation at the houses (Evans and Abrams 1988). These may have been the special tools for scraping down the maguey leaves *(pencas)* for fiber. Parsons and Parsons (1985:9, 17) describe modern maguey farmers doing this scraping with a metal tool of a shape similar to the basalt scrapers (it is made from a section of automobile leaf spring), and experimental use of the basalt implement in the scraping process showed it to be effective.

Maguey fiber was made into a variety of products, including woven textiles. We have no direct evidence of weaving at Cihuatecpan (the loom and other tools were all of perishable materials), but ceramic spindle whorls are common. Of the 122 spindle whorls recovered from excavation in houses, 62 (barely over half) were large, a type Linn assumed was used for maguey fiber (Linn 1934:128, cited in Parsons 1972:61). Parsons and Parsons (1985:35c) substantiated this assumption in their ethnographic studies of modern maguey farmers and found a direct relationship between spindle whorl size and thread thickness. The large spindle whorls at Cihuatecpan are yet another indicator of maguey exploitation. Fiber processing and sap processing involved tasks that lent themselves to a gender-based division of labor, and the nuclear and joint household types known from the Aztec period are well suited to these cooperative efforts (Evans 1990, n.d.).

When Cihuatecpan and other dispersed villages were at their most extensive, their terraces covered the valley's piedmont and they held half the population of the Teotihuacan Valley (more than 60,000 people of about 130,000; Evans 1980:155). The villages were patterned so as to take maximum economic advantage of a semiarid area, marginal for grain crop cultivation but ideal for xerophytic plants: nopal (*Opuntia* spp.) and, particularly, maguey (*Agave* spp.).

Maguey Farming and Settlement Patterning

It should be noted that the ecological balance of the settlement strategy of these villages depended on the active maintenance of the terrace systems where maguey and other crops grew, and this constraint demanded that the farmer live close to the crops, to tend the terraces (Sanders 1965) as well as the plants (Parsons and Parsons 1985, 1987; Drennan 1988:285 cites labor requirements in general as a determinant of dispersed settlement in ancient Mesoamerica). A further benefit of this pattern was the enrichment of the soil through use of household waste; in a setting lacking large draft animals and their fertilizing potential, the value of night soil and other biodegradable organic material applied to house gardens is not inconsiderable (Palerm 1955:29).

These constraints produced the dispersed house-lot pattern. The map of Teotihuacan Valley settlement (fig. 5-2) shows how these villages extended along the valley's piedmont in a continuous sweep of settlement. This dispersed settlement type demonstrates the crucial relationship between the nature of the environment and the adaptive patterns of culture. The lands on which these people had settled owe their high productivity to each household's commitment to the many facets of maguey exploitation, because the terraces were the *calmil* and *milpa* plots of the village. On the alluvial plain were the more fertile fields, but control of these was divided among various nobles and other landholders and civil and religious functionaries. With the probable exception of Cihuatecpan's headman (who may have had some alluvial plain land dedicated to the support of his office), the beneficiaries of alluvial plain productivity would have been urban dwellers, not peasant villagers.

Land and Landholding in Aztec Times

In order to reconstruct how Cihuatecpan would have used its landscape, certain Aztec categories of landholding must be understood. The Aztecs recognized a multiplicity of named types of landholdings (Gibson 1964: 257, 267) and in table 5-1 those relevant to the upper Teotihuacan Valley are listed. The sum of a city-state's land, the *altepetlalli* (Gibson 1964:267), would encompass the communal lands of the polity's tribute-paying peasants *(calpullalli)* and the lands set aside for support of its political, social, and religious institutions (such as *tlatocatlalli* for the ruler, *tecpantlalli* for the barrio and village headmen,

Table 5-1. Cost/Benefit Analysis of Land and Landholding

	Categories of Land and Landholding					
	Altepetlalli (e.g., of Otumba city-state)					
	Calpulli Land (e.g., of Cihuatecpan)					
	Tlalmil					
Categories of Landholders	Calmil	Milpa	Tecpantlalli	Teotlalli	Tlatocatlalli	Pillalli
City-state ruler					+	
Village headman	+	+	+			
Landed commoners	− and +	− and +	−	−		possible +
Headman's workers	possible −	possible −	−			possible +

Note: Minus signs (−) represent labor cost and plus signs (+) represent a benefit to be derived from control over the food resource being produced.

and *teotlalli* for the temples and priests). Some lands (*pillalli, tecuhtlalli*) supported particular nobles and their heirs, but in these cases the workers on the land did not owe tribute to the local lord, and such lands would not be part of the *altepetlalli*.

All citizens of a particular city-state would be expected to participate in numerous public service projects for various levels of government, from tending to the needs of their own village to occasional household service in the palaces of the Acolhua confederation capital at Texcoco. Table 5-1 presents a breakdown of labor cost (minus signs) and benefit from control over the food resource being produced (plus signs) for various types of land, by relevant social categories. The behavioral model expressed by table 5-1 assumes that the city-state ruler drew food from plots dedicated to his use and that *calpulli* land would have provisioned the villagers and their headman. In fact, it is unclear from documentary sources whether the food required as tribute in the Codex Mendoza would have come from the *calpullalli* or from special plots,[4] but food tribute for other offices clearly was produced on special plots probably located on the alluvial plain.[5] This practice had the consequence of reducing the tax bite as it applied to productivity of *calpullalli* land itself. Such land was still taxed in certain kinds of "produce" (for example, woven cloth from fibers harvested from maguey plants), but the actual food produced from the *calpullalli*'s *calmil* and *milpa* plots was probably under the control of the producers and their headman, to use or to trade.

Decline of Village Life

The productivity of village land began to decline not long after the Spanish conquest of Mexico in 1521; throughout the sixteenth century, Cihuatecpan and many similar villages diminished in size. The precipitous declines in population size in many parts of the New World resulted from the deadly combination of European diseases and European greed. So diminished was the rural population of the Basin of Mexico that the colonial government took steps to bring the survivors together, forcing them to abandon their villages and take up residence in congregated communities. Cihuatecpan's turn came in 1603, when the Orden de Congregación of October 3 called for San Lucas Siguatecpan to be abandoned; the remnant group of villagers was ordered to relocate in nearby Ahuatepec (Archivo General de la Nación 1603:34R).

After the mandated abandonment of San Lucas Cihuatecpan, some sporadic occupation continued, as evidenced by Colonial period sherds. These were concentrated around the south slope of the cerro, and a disproportionately large number came from Structure 6 (82 percent of the glazed sherds, which contained only 38 percent of all sherds from 1984 research operations; Evans 1988:37–38, 43), indicating that it continued to be occupied after most of the other houses had been abandoned. Cihuatecpan, the dispersed village of the Aztec period, had ceased to be, and the land around Cerro San Lucas probably became grazing land, like so much of the upper Teotihuacan Valley. The upkeep of the terrace systems would no longer be necessary or feasible, and erosion proceeded unchecked. The ruins of Cihuatecpan's houses were covered by water- and wind-borne soil, and the village was forgotten until the 1960s, when the site was mapped by the Teotihuacan Valley Project.

ENVIRONMENT AND AGRICULTURAL PRODUCTIVITY

Remnants of the ancient terrace system on Cerro San Lucas still exist, judging from the spatial relation of visible eroded terrace borders to Aztec period house mounds (though much of the upper south slope has been recently reworked with the help of bulldozers, leading to predictable damage to house mounds). The construction and maintenance of the ancient terrace system (which presumably was established at the same time as the village) represents the imposition over the landscape of a sophisticated artificial ecosystem, designed to optimize adaptation to the challenging climatic conditions of low rainfall (500 millimeters per year), cool climate (frost possible between October and April), and high altitude (occupation zone ranges from 2,400 to 2,500 meters above sea level).

Cihuatecpan is in the upper reaches of the Teotihuacan Valley, and the valley as a whole has a cool, arid climate with some variation in temperature and rainfall, depending on the topography. The valley covers about 600 square kilometers and is defined by the drainage of a chain of hills (the highest hill, Cerro Gordo, is 3,050 meters above sea level) surrounding a northeast to southwest expanse of alluvial plain that ends at the ancient shoreline of Lake Texcoco (at 2,240 meters above sea level).

Rainfall in the Teotihuacan Valley is strongly seasonal (from late spring into fall) and follows a marked diurnal pattern, with torrential showers falling virtually every summer afternoon. The predictability of this daily pattern definitely works to the advantage of villagers living close to agricultural terraces, inasmuch as situational channeling of runoff onto the terraces increases crop security.

The timing of yearly seasons is somewhat less predictable, and the onset and duration of the rainy season are chronic matters of concern; four years out of ten, rainfall is below 600 millimeters (Nichols 1987:604–05). The most critical time for the security of a good yearly harvest is spring, when late frost could kill off seedlings. However, when seeds are deeply planted and then irrigated, this risk is considerably diminished.

Perhaps because of the Teotihuacan Valley's climatic marginality for grain cultivation, significant settlement there was established only a few hundred years before Christ, when exploitation of the permanent springs at Teotihuacan for agricultural use probably began, and irrigation ameliorated some of the effects of the climate in the lower valley.

In contrast, the water supply for the settlements in the upper valley in prehispanic times depended on the careful husbanding of runoff from occasional small springs and from rainfall. Water was stored in pondlike reservoirs now called *jagueys* (at Cihuatecpan, one *jaguey* was located on the southern slope [described in Evans and Abrams 1988:84–86], and runoff was channeled over terraces and into floodwater irrigation fields). In the lower Teotihuacan Valley, climatic exigencies (unseasonal frost, delayed rainfall) could to some extent be lessened by judicious use of permanent water sources, but in the upper valley fewer buffers existed to protect against the loss of seed crops from drought.

Terrace systems solved these problems, providing better soil conditions for plant germination and growth plus the added productive value of maguey and nopal. The terraces harness erosion, because drainage of rainfall runoff through the system brings soil in solution as well as water itself onto the terrace surfaces, and in time crop security is enhanced by the combination of deeper soils and more deeply held moisture (Donkin 1979; Sanders 1965:39–44). The dynamic elements of the drainage process are used to create a particular cultivation system, and although agricultural conditions in the piedmont of the upper Teotihuacan Valley were marginal for maize and other grains, intensive labor inputs created a productive landscape by reducing risks and by diversifying the resource base, with the use of maguey and nopal edge plantings on the terraces.

Fig. 5-3 shows the layout of the village around the cerro; the area between the dotted lines encompasses the cerro's slopes and delineates the survey zone of about 330 hectares. Below the lower survey limit the alluvial plain begins, and above the upper limit the slope becomes much steeper and is now covered with thorny shrubs. Habitation was continuous, but clear clustering of house mounds is demonstrated on the southern slope, the area where the roads and major drainage channels intersect, and the house mounds here are larger and denser than they are elsewhere on the cerro. The desirability of this south slope results from the intersection of a variety of environmental and cultural features: greatest access to sunshine and to drainage from the caldera of the cerro was important but so was proximity to the trade route used by travelers between Otumba and the gulf coast (a route that followed or paralleled one of the present roads).

The terrace system around Cerro San Lucas encompassed two different kinds of horticultural strategies: first, grains would be grown on the flat areas and maguey or nopal on the embankments, and this procedure would constitute *milpa* farming; second, the terrace surfaces close to the houses would be used for kitchen gardens *(calmilli)*, where tomatoes, chiles, and other vegetables as well as herbs and flowers and a small amount of corn (for edible green corn and as a backup for the next year's seeds) would have been grown for household use. This pattern may have resembled that of the modern community of Eloxochitlan, a Totonac village with dispersed house lots in Puebla. There, *calmil* plots averaged .5 hectare per house, and each family also farmed a larger *milpa* plot farther away (Bray 1972:910; Palerm 1955:29).

If we were to map out an area on Cerro San Lucas that is adjacent to the house lots and encompasses .5 hectare per house, about 101.5 hectares of the village's 330 hectares of sloping land would be used; about 228.5 hectares of sloping land remain on the cerro for *milpa*. Thus the sloping land on Cerro San Lucas permits each household a small adjacent kitchen garden and a larger, more distant farm field.

The productivity of *milpa* land is in part based on the average harvest of maize or other grain. The Teotihuacan Valley as a whole was an important grain-producing area in Aztec times, probably capable of supporting its own population (Sanders 1970b:442; Evans 1980:168) and of exporting maize and other grains as part of the tribute requirement documented in the Codex Mendoza. The lion's share of the valley's grain production came from the permanently irrigated lower valley where yields may have equaled as much as 3,000 kilograms per hectare, but on some parts of the piedmont maize yields might have been as little as 300 kilograms per hectare. Around Cerro San Lucas, the average annual yield would have been about 400 kilograms per hectare (Sanders 1976:144). Productivity here was somewhat lower than that of other piedmont zones, because the runoff available was dispersed all around the cone of the cerro rather than being more concentrated (the latter effect characterizes the topography of the Patlachique and Malpais ranges and their piedmont zones).

While the valley as a whole was capable of supporting its regional population, land varied greatly in quality, and if the piedmont dwellers had to depend only on grain crops from their terraces for local caloric needs, they could not support themselves very well. Williams used ethnohistoric documents to study the productivity of terraced villages in the piedmont zone around the Texcoco plain, just south of the Teotihuacan Valley. Based on grain productivity alone, "extreme population pressure clearly existed" in this area (Williams 1989:720). Similarly, my calculations for Cihuatecpan's grain productivity determined that "Cerro San Lucas could not fill the food needs of the village" (Evans 1988:23). But both sets of calculations are limited to grain, and maguey sap may have constituted another major food crop, one that greatly expanded the nutritional base of maguey farmers.

Fresh sap, called *aguamiel*, is very watery with a pleasant sweet-tart flavor. Within a day the texture becomes viscous as the sap begins to ferment into *pulque*. But when the sap is at its freshest it no doubt provided the standard beverage for maguey farmers and their families, and it substituted for fresh water often lacking in the piedmont zones of the central highlands. This use of *aguamiel* illustrates Sauer's comment; for these piedmont farmers, maguey is a source of potable (and nutritious) drink, without which habitation would be impossible.

Studies of maguey terrace agriculturalists in Orizabita (an Otomi village in the Mesquital area, north of the Basin of Mexico; Parsons and Parsons 1985, 1987) and in the Teotihuacan Valley itself (Sanders 1965) have documented strategies and yields in maguey farming, and from these the productivity of the Cerro San Lucas area for the Cihauatecpanecos can be estimated. It should be noted that these ethnographic studies calculated the productivity for maguey in mixed-farming contexts, interplanted with maize and other grains.

In table 5-2, values are calculated for the area and productivity of land in use in and around Cerro San Lucas to estimate the ability of the land to support the Cihuatecpanecos at a subsistence level. Here the combination of calories from these staples, maguey and grains, is assumed to make up most of the diet. (As Sanders [1976:109] has pointed out, maize makes up 80 percent of the diet in many peasant communities of Mexico and Guatemala today.)

Maguey contributes significantly to the diet, and this contribution is calculated as the caloric value of *aguamiel, pulque,* or the solid sugar products derivable from *aguamiel* (the caloric yield per plant remains generally the same, no matter how the sap is processed). To understand the productivity, in calories, of maguey in mixed maguey and grain cultivation, we use observations from ethnographic studies.

The maguey holdings in use and their productivity are estimated from these observations:

- A single household "cultivates a few hundred maguey plants, and produces a few dozen liters of pulque a day" (Parsons and Parsons 1987:83).
- For steady production of about 5 liters per day, about 40 plants (including nonproducers) are needed (Parsons and Parsons 1987:84; Sanders 1965:45); therefore, average yield is roughly 0.125 liters per day per plant, or 45.6 liters per year per plant.

Table 5-2. Area and Productivity of Agricultural Land, Cerro San Lucas

	Categories of Land and Landholding					
	Altepetlalli (e.g., of Otumba city-state)					
	Calpulli Land (e.g., of Cihuatecpan)					
	Tlalmil					
Production Factors	*Calmil*	*Milpa*	*Tecpantlalli*	*Teotlalli*	*Tlatocatlalli*	*Pillalli*
Type of land	piedmont			alluvial plain		
Crops cultivated	maguey and vegetables	maguey and grain		grain and some maguey		
Caloric value/ha/yr	2,087,650 (from maguey)	3,527,650 (from maguey and grain)		4,121,913		
Area in use (ha)	101.5	228.5	11[a]	yes[b]	yes[b]	possible[b]
Caloric yield, in millions of kcal/yr	211.89647	806.0802				
Individuals supported (at 2,500 kcal/day or 912,500 kcal/yr)	232	883	50[a]	?	?	?
Total	1,115					

[a] The *tecpan* probably housed about 25 people and offered hospitality and charity to many others. If we assume the *tecpan* supported 50 people a day, each needing an average of 2,500 kcal, then the total caloric requirement for the year would be 45,625,000, and 11 ha on the alluvial plain would be required to provide it (each hectare supporting 4.5 people at 912,500 kcal/year each).
[b] The area of these plots and the village's investment of time are not known. The caloric value would be calculated as in note a.

MAGUEY TERRACE AGRICULTURE

- Average sap production per year in mixed holdings equals 4,855 liters per hectare (Parsons and Parsons 1987:88); and 4,855 divided by 45.6 liters per plant equals 106 plants, therefore planting density per hectare equals 106 plants, producing 13.25 liters per hectare per day.

To apply these values to Aztec period Cihuatecpan, we assume a mixed planting strategy and land use area as our basis for extrapolating productivity. We assume that each household at Cihuatecpan cultivated a *calmil* plot and a *milpa* plot; the areas of these (according to this model) are assumed to be (respectively) .5 hectare and 1.1 hectares, based on 203 houses and 330 hectares available. *Milpa* productivity may be calculated as a sum of maize and maguey productivity in kilocalories (caloric values are from Woot-Tsuen and Flores 1961). Note that the calculations use probable values for interplanted crops, and the values per unit of area are understood to be cumulative rather than mutually exclusive:

If maize (400 kilograms per hectare; 3,600 kilocalories per kilogram) = 1,440,000 kilocalories per hectare, and maguey (4,855 liters per hectare; 430 kilocalories per liter) = 2,087,650 kilocalories per hectare, then *milpa* (interplanted maize and maguey) = 3,527,650 kilocalories per hectare,

a daily yield of 9,665 kilocalories and 13.3 liters of potable beverage per hectare. Each household's *milpa* = 1.1 hectare = 3,880,415 kilocalories per year and 5,340.5 liters of potable liquid. Daily yields are 10,631 kilocalories and 14.6 liters per average *milpa*.

Productivity of *calmil* plots is more difficult to calculate. While the important nutritional contribution of crops grown in these kitchen gardens is significant, their caloric value is typically low, and the zone covered by the house and circumhouse activity areas also reduces this value. In this model, the caloric value of such crops will not be estimated. The caloric value derived from maguey planted in the *calmil* area of .5 hectare as a terrace border and among other crops will stand as the basic caloric productivity of the *calmil*. The same general density of planting will be assumed, because the houses were built on the terrace beds and not astride the borders where many of the maguey plants were established. If maguey produces 2,087,650 kilocalories per hectare, then the .5 hectare *calmil* plot would yield 1,043,825 kilocalories per year and 2,427.5 liters of potable liquid. Daily yields are 2,860 kilocalories and 6.7 liters.

Thus, the total productivity (grain plus maguey sap) of each household's *milpa* and *calmil* plots (together, 1.6 hectares) is 4,924,240 kilocalories per year, including 7,768 liters of potable liquid. Daily yields are 13,491 kilocalories per day, including 21.3 liters of potable liquid. That this is calorically sufficient to support at least five or six people at the subsistence level is impressive, because that number of people was probably the average household size, based on ethnohistoric evidence from the Colonial period and inferred from the sizes of excavated houses (Carrasco 1976; Evans 1989; Harvey 1985). Cihuatecpan would have had a valuable kind of economic self-sufficiency, buffering the village from the potentially devastating consequences of permanent reliance on the city markets for food. Thus, the slopes of Cerro San Lucas were sufficient to provide for the population of Cihuatecpan, given a mixed strategy of maguey and grain cultivation.

In addition to the calories, and in some sense equally important in permitting successful adaptation to this challenging landscape, is the yield of potable liquid. Each person would be provided with about 3 liters of fresh maguey sap a day, sufficient to substitute for much of the basic water requirement. A survival ration of water for drinking is 1 liter per day per adult male in a neutral (stress-free) physiological state (Davidson et al. 1979:542). Matheny's (1978:204) research on present-day water use in the Edzna Valley (Yucatan peninsula, Mexico) indicates an average of 12.5 liters per day per person, including frequent bathing but excluding watering animals and washing clothes. If Cihuatecpanecos used much of their maguey sap harvest as a fresh beverage, their general need for water would be much reduced. The capacity of the *jaguey* at Cihuatecpan can be estimated from its projected depth and surface area and would have been about 260,000 liters. If we assume a village population of about 1,000, it is clear that one *jaguey* would have been insufficient to serve all the village through the dry season, and furthermore, *jaguey* water can easily become too polluted to drink (Siliceo 1922:190–92). In providing a year-round reliable potable beverage, maguey permitted settlement in areas lacking permanent water supplies.

It should be noted that the system of local land use modeled here is not expandable, and little capability exists for creating food surpluses by further intensification. However, it should also be noted that the values of maguey and grains produced on Cerro San Lucas are revealed in an artificially depressed state when considered only as caloric inputs. The market value for *pulque* and other maguey products may have been so high that it was economically advantageous for the Cihuatecpanecos to trade *pulque* for maize and other grains grown elsewhere in the Basin of Mexico.

OTHER COSTS AND BENEFITS

Having established this very basic relationship between the villagers and their local food supply, the model must now be adapted to account for village life in the larger cultural context. The ability of *aguamiel* and maguey hearts to sustain life in starvation times no doubt served Cihuatecpan well in the famine years of the mid-fifteenth century, but in normal times maguey products could be traded for other goods or food, worth much more than the basic caloric value of the maguey plant.

Sap. The average household production of 7,768 liters per year of *aguamiel* was discussed above as a potable beverage with a caloric content similar to *pulque* (which is similar to that of beer). What was not drunk as fresh sap in the village was either permitted to ferment, for local use or trade, or was processed into solid sugar, which has the advantage of long-term storage stability. At 16 liters of *aguamiel* per 1 kilogram of sugar, a household could produce up to 485.5 kilograms of sugar a year and probably devoted some of its *aguamiel* to this purpose. The details of this process are unclear, but it may well have required considerable valuable fuel.

Food Value of Other Parts of the Maguey. The maguey heart can be chewed for the sap, but the fiber is indigestible. Maguey quids are common at Early Formative campsites in Mesoamerica, and to this day maguey heart is known as a reliable buffer against starvation. It provides few calories for the effort of mastication, however, and this use would preclude the many other ways of exploiting the plant; for these reasons we assume that the quids were chewed only in times of extreme hardship. When any other food was available, the plant would serve its multiple purposes, and the heart of maguey might be processed into luxuriously silky thread; when the famine years came, the heart of maguey kept the edge off raw hunger and provided a few calories, some phosphorous, iron, and B and C vitamins.

Fibers. Fibers from the heart of the maguey, but even more importantly from maguey *pencas* (leaves), were vital to the village economy. At Cihuatecpan evidence of fiber production is found in every house: basalt maguey scrapers (to strip down the fibers from the *pencas*) and ceramic spindle whorls document two phases of the complex fiber and cloth production process. In addition to producing thread and standard woven mantles, the households probably also made rope, nets, and baskets as these were needed by the family or were marketable. Woven textiles constituted a sort of currency in the three exchange systems (tribute, long-distance trade, and marketplace), and weaving skills were highly valued in women because they were the household-level producers of massive numbers of textiles. To fulfill its share of the Codex Mendoza tribute alone, each Cihuatecpan household had to produce four long woven mantles a year.

Other Uses. Other uses of the plant include medicines derived from the sap and the probable use of *pencas* for construction material and for animal fodder (if the *pencas* were immature). The dead plant was burned for fuel, and its ashes would have been used as fertilizer.

The maguey plant was clearly a mainstay of the village economy, and in essential value it outweighed nopal, the other common xerophytic "orchard" plant in the Teotihuacan Valley. The foods nopal produces (cactus paddles as a green vegetable and tuna, the pink, heart-shaped fruit) are important for rounding out the diet (Woot-Tsuen and Flores 1961:33, 37, 59, 62) but do not sustain life. Nor is nopal suited to a wide range of nondietary uses, as is maguey.

Nopal did, however, produce a very valuable crop, the cochineal insect. The processing of the insect into dye can be easily managed with implements found around the Aztec house, and so valuable was the dye that even the rinse waters of the implements were used to impart color to fibers. Because specialized tools apparently were not used, and residues of the dye were removed in the dye process, archaeological evidence for this industry is lacking (or indistinguishable from evidence of more ordinary pursuits). But early colonial documents refer to the strength of the industry in this general area and to its importance (second only to gold) in the colonial export economy (Gibson 1964:354). Cochineal was also highly valued in Aztec times and would have been a natural sideline of textile production at Cihuatecpan.

CONCLUSION

The cultivation of maguey and other xerophytic plants in the central highlands of Mexico transformed the sloping piedmont zone into a productive landscape several centuries before the Spanish conquest. Maguey gave peasant farmers a stable source of raw materials for craft production and provided basic caloric and liquid requirements. The versatility of maguey's uses permitted the settlement and exploitation of a broad area, contributing substantially to the demographic and economic strength of the Aztec period.

These conclusions are drawn from information compiled from archaeological, ethnohistoric, and ethnographic sources. The Aztec period village of Cihuatecpan is the case study, the basis for reconstruction of the agricultural resource base. Ethnohistoric documentation of maguey production and ethnographic studies on maguey, as well as documentation of settlement and agricultural patterns, permit us to estimate the agricultural strategies in use at Cihuatecpan and maguey's importance there and in countless other peasant villages. The interplay of maguey farming and settlement of the piedmont lands resulted in Aztec period population levels in the Teotihuacan Valley that were nearly as high as those of the Classic period a thousand years before. What exploitation of the springs at Teotihuacan had done to expand the productivity of the valley's alluvial plain during the Classic period, exploitation of maguey did for the slopes a millennium later.

ACKNOWLEDGMENTS

Excavations at Cihuatecpan were supported by NSF grant BNS 8317830 and study of materials facilitated by NSF grant BNS 8519834. This study was first published in *Latin American Antiquity*. I appreciate comments on previous drafts made by Thomas W. Killion, William T. Sanders, Michael E. Smith, B. L. Turner II, David Webster, and several other reviewers.

NOTES

1. On the use of the term *Aztec:* at the time of the Spanish conquest, most of the peoples of the Basin of Mexico were Nahuatl speakers and claimed at least partial descent from migrant groups from the north who arrived sometime after A.D. 1000. These migrants were ethnically diverse but had interwoven histories and origin myths, and a major feature they had in common was the idea of a legendary origin place, Aztlan. These groups came to be called *Aztecs* because of this idea, and even though that name is most closely associated with the Mexica of Tenochtitlan, it applies by and large to most of the peoples of the Basin of Mexico from A.D. 1150 to 1521, an era called the Aztec period. This reconstruction from documentary sources of the Early Colonial period (most important for the Teotihuacan Valley are the Codex Xolotl and the *Obras Historicas* of Ixtlilxochitl) has been substantiated by the archaeological evidence gathered by the Basin of Mexico Project (Sanders, Parsons, and Santley 1979) and by Instituto Nacional de Antropología e História research (e.g., Vega Sosa 1979).

2. *Milpa,* in this context, means an agricultural plot located at some distance from the house, *"en el campo"* (Simeon 1984:276), a pattern found in many parts of Latin America today.

3. Generalizations concerning the florescence of Tula and contemporaneous sites are based on ethnohistoric and archaeological evidence, such as that reported by Cobean and Mastache (1989), Cobean et al. (1981), Diehl (1981, 1983), Healan, Cobean, and Diehl (1989), and Kelley (1971). Armillas (1969), using climatic data from other parts of the world, hypothesized that a shift in the climate of this semiarid area greatly reduced its arable land.

4. Pages 21 and 22 of the Codex Mendoza list a number of Teotihuacan Valley towns, including Otumba, Ahuatepec, and Axapusco, and the tributes

required are bins of maize, beans, chia, and amaranth (Barlow 1949:71), fabric and clothing, war dresses, and shields. Motolinia describes the tribute owed by Teotihuacan Valley towns to Texcoco as including lime, stone, firewood and firebrands, lumber, and peasant labor (Motolinia 1971:394–95).

5. The settlement pattern of the valley strongly favors the piedmont zone right above the alluvial plain, permitting conservation of the best land for agriculture and easy access to it. Gibson notes that the "lands worked in common for the benefit of local caciques and principales are described as adjoining the towns and worked by 'all the people together' for periods of two or three hours per day. . . . they were 'common' lands for the support of the offices of rule" (1964:259).

PART II

Artifact Distributions and the Organization of Prehistoric Agriculture

EVIDENCE FROM LOWLAND MESOAMERICA

CHAPTER 6

THOMAS W. KILLION

Residential Ethnoarchaeology and Ancient Site Structure

CONTEMPORARY FARMING AND PREHISTORIC
SETTLEMENT AGRICULTURE AT MATACAPAN,
VERACRUZ, MEXICO

SETTLEMENT AND AGRICULTURAL SPACE coalesce in the humid tropics. Buildings, open areas, gardens, and fields merge, obscuring boundaries and the managed character of a densely populated landscape. The spatial integration of cultivated and residential areas within settlements contrasts sharply with the segregation of settlement and agriculture on landscapes outside the humid tropics. In more arid or temperate settings, towns and villages—the "built" component of the farmer's landscape—and agricultural space are more clearly defined. Not only do the residential and cultivated portions of the local environment tend to overlap in the humid tropics but the number of components in the system are also more diverse. A wide variety of orchards, gardens, and agricultural fields interdigitates with settlement space in a way rarely observed in drier and cooler regions. The material outcome of this mixture is the garden residence, a fundamental feature of tropical living.

The distinctions between humid tropical and temperate or arid settlement and agriculture have important implications for understanding the archaeological record and the evolution of prehistoric agricultural systems. The location of agricultural fields and the manner in which they are cultivated greatly affects the organization of household labor and the role of the residential lot as a center for household farming operations. The long-term use of residential areas produces site structural patterns that provide a diagnostic link to variation in local agricultural production and the diversity of peasant household farming activities.

This chapter examines the organization of production in a humid tropical agricultural system from the site structural perspective of the garden residence. Following an introduction to residential site structure based on examples of settlement systems located in the New World tropics, I examine variation in peasant house-lot site structure in the Sierra de los Tuxtlas of southern Veracruz, Mexico. A model of residential site structure is presented relevant to agricultural settlements in similar environments. The model focuses on the long-term management of habitation structures, activity areas, refuse zones, and cultivated spaces as fundamental determinants to the formation of the archaeological record in humid tropical settings. Data are then presented that suggest that basic elements of residential site structure are correlated with the location of agricultural fields and the intensity of cultivation on them. In the final portion of the chapter I apply the model developed from contemporary contexts to the archaeological record of settlement at the Classic period regional center of Matacapan, located in the center of the Tuxtlas region. The site structural approach, based on an understanding of contemporary settlement and agriculture, yields insight into community structure and the use of residential space at prehistoric Matacapan. The proposed blend of agricultural and settlement space at Matacapan represents the first step in developing a functional understanding of prehistoric settlement in the humid tropics.

RESIDENTIAL SITE STRUCTURE IN THE HUMID TROPICS

Residence, garden, and field are the essential elements of settlement throughout the lowlands of tropical America. At the center of lowland settlement is the garden residence. The spatial structure of the residential area results from the patterned use of space by household members. The garden residence itself is made up of living quarters, clear activity spaces, refuse scatters, and vegetated areas. Different use areas surrounding the residence are distinguished by plant cover, refuse, clear space, and structures, all of which are enclosed within the borders of the household's residential compound. Use areas around the residence expand and contract in size due to family demography, the economic focus of the household, and the availability of space within settlements. The spatial juxtaposition of gardens, middens, clear areas, and houses often crosscuts simple household groupings, yet within agricultural settlements these use areas are the standard and redundant elements of residential space.

The spatial components of the garden residence are found in a wide range of demographic contexts in the New World tropics. Ethnographic documentation from the Amazon Basin, for example, includes Bergman's (1980) description of Shipibo garden residential settlement. Shipibo houses are surrounded by dooryards made up of clear areas, debris-laden space, and adjacent gardens and orchard areas. Denevan and Schwerin (1978) describe a similar arrangement among the Karinya, who inhabit the seasonally inundated lands of the Venezuelan Llanos. Outside the borders of the house and its surrounding clear area, Karinya house lots include a garden whose growth is enhanced by household refuse and night-soil deposition. Information concerning the site structural elements of the garden residence under conditions of low population density elsewhere in the lowlands of South America are found in Beckerman (1983), Gregor (1977), Harner (1972), Hiraoka (1986), and Johnson (1983). Additional examples of the garden residence in Central America and the Caribbean can be found in Covich and Nickerson (1966), Kimber (1966, 1973), and Wagner (1958).

In the more heavily populated lowland environments of eastern Mesoamerica (Mexico, Guatemala, and Honduras) descriptions of the garden residence are found in the work of Bullard (1952), Hester (1954), Lundell (1933), Steggerda (1941), and Wauchope (1938). In Mexico, residential associations of gardens, clear areas, and habitational structures have been described for subsistence agriculturalists in the Tarascan area (West 1947), among the highland Maya at Zinacantan (Vogt 1969), and in the Tajin Totonac territory of Gulf Coast Veracruz (Kelly and Palerm 1952).

Early chronicles and other ethnohistoric records (Anghiera 1912; Hellmuth 1977; Roys, Scholes, and Adams 1940, 1959; Scholes and

Roys 1968) provide a wealth of details concerning the spatial structure of the tropical garden residence. Numerous scholars have argued for its great antiquity and potential value in understanding the evolution of tropical agriculture and settlement (Anderson 1952; Harris 1972; Lathrap 1977). The wide distribution and apparent antiquity of the garden residence underline the importance of developing an archaeological understanding of this ubiquitous element of tropical living.

Alcorn's study of Huastec Maya ethnobotany (1984) in San Luis Potosi and northern Veracruz, Mexico, contains detailed information on garden residential site structure and the use of space in the extramural areas surrounding Huasteca houses. Her analysis of Huastec spatial perceptions suggests a strong link between how people think about space and the material outcome of the regime of daily activity. Alcorn develops a model of the Huastec garden residence and its place in local patterns of settlement and agriculture. At the center of this model is the household's residential structure surrounded by a clear area of bare earth and peripheral vegetation. The bare zone immediately adjacent to the house is the outdoor activity area, referred to as the *eleeb*. Huastec perceptions of the areas around the residence correspond to the material contents and spatial compartmentalization of the garden residence outlined above. Alcorn relates how the Huastec farmer thinks about residential space in his own words: "*Wal eleeb* (literally face/edge/eye of *eleeb*) specifically refers to the vegetated area bordering the clear *eleeb* . . . 'The edge of the dooryard where one stops sweeping is for the animals, it is dangerous to walk there. There are thorns, rubbish, snakes. Here one can put one's animals. It is "bush," it is for animals not for people. It gets muddy here, we throw rubbish here, it is shady here'" (1984:117).

The pattern that emerges from Alcorn's description of the Huastec garden residence includes the same basic components of site structure consistently repeated throughout the tropical lowlands of Mesoamerica. The residential area is centered on habitational structures. Surrounding the structures of the "built environment" is a clear area, maintained free of debris for the economic, social, leisure, recreational, and ritual activities of the household. The clear area in turn is surrounded by a zone of debris generated by the household and finally by rings of cultivated and wild vegetation. This basic structure forms the "playing surface" for the daily activities of the household. These activities have material consequences for residential site structure and ultimately have implications for our reading of the archaeological record of tropical lowland settlement.

RESIDENTIAL SITE STRUCTURE IN THE TUXTLAS REGION

Detailed studies of residential site structure recently completed in the Sierra de los Tuxtlas of southern Veracruz, Mexico, have provided a wealth of new information on peasant agriculture and household production in a tropical lowland setting of Mesoamerica (Arnold 1986; Killion 1987a). The Tuxtlas (fig. 6-1) are a small group of volcanic mountains rising 300 meters or more above the otherwise flat gulf coastal plain of southern Veracruz. The sierra is one of the wettest regions anywhere in Mexico or Central America and less than 100 years ago supported dense stands of tropical rain forest (Gomez-Pompa 1973). Today, most peasant farmers in the Tuxtlas practice a mixture of subsistence agriculture, cash cropping, and wage labor. Commercial agriculture (sugarcane, banana, pineapple, and tobacco) and extensive cattle grazing by wealthy nonpeasant families play a central economic role in the Tuxtlas. Plantation crops monopolize the most valuable flat agricultural land in the region while subsistence agriculture and cattle production compete on the areally more extensive slope lands of the Tuxtlas (Killion 1987a).

Residential site structures among peasant *(campesino)* communities in the Tuxtlas is closely linked to activities conducted within the garden residence that are normally integrated with agricultural activities conducted outside the residential lot. A relationship can be shown to exist, then, between site structural variation *within* the house lot and the organization of agricultural activity *outside* the lot. Specifically, the intensity of cultivation on agricultural plots located at different distances from the residence is shown to have a measurable effect on the size of use areas within the garden residence.

The organizational and spatial focus of most tropical farming systems, including the peasant farming system found in the Tuxtlas, is the household's residential structures and surrounding extramural areas.

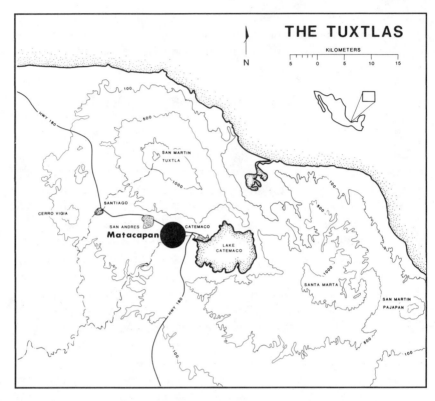

Figure 6-1. The Tuxtlas Region Showing the Area of Contemporary Ethnoarchaeological Studies and the Middle Classic Center of Matacapan

The house-lot model shown schematically in figure 6-2 is an ideal representation of house-lot site structure. The house-lot model is made up of a number of spatial components. These components include the structural core, a clear area of debris-free space surrounding the core, an intermediate area of fairly concentrated refuse enclosing the clear area, and a peripheral garden of mixed vegetation and debris. The model specifies that the use of space within domestic contexts is related to the daily management of the residential area as well as a variety of activities that take place outside the residential locus. The regular management of debris-free and debris-laden space within the lot creates a

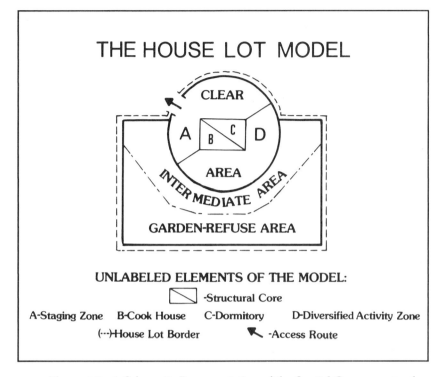

Figure 6-2. A Schematic Representation of the Spatial Components of the House-Lot Model

distinctive set of spatial and material patterns around residential structures.

Deal (1983) and Hayden and Cannon (1982) have presented a model of Tzeltal Maya residential site structure that closely corresponds to the house-lot model presented here. These authors discuss elements of peasant residential site structure in the context of pottery disposal and the material correlates of status. Santley (this volume) discusses differences in the two models and their implications for the interpretation of Formative period site records. The present discussion focuses on diagnosing features of local agricultural production based on an examination of the use of space within the residential lot.

The Structural Core

At the center of the Tuxteco house lot is the *structural core* consisting of the primary living structures, storage buildings, and other built facilities that shelter household members and their belongings. The size of the structural core in the Tuxtlas region varies closely with the number of people inhabiting the house lot (Killion 1987a), following the general relationship proposed by Naroll (1962). Sleeping, resting, cooking, food preparation, and storage are the primary activities that take place within the roofed-over space of the primary residential structures. Activities associated with cooking and food preparation and those associated with storage and sleeping are generally segregated within the core. A common structural response to the spatial opposition of these activities in the Tuxtlas region is the construction of a separate *cocina*, or cook house, and *casa*, or dormitory.

The living space within the structural core also includes porches and space under eaves where people can work, play, and rest. The exterior walls of buildings are used for the storage of numerous items used inside and outside the lot. Functional household items and provisionally discarded debris accumulate in this interstitial portion of the core. Provisional discard refers to placement of broken or otherwise nonfunctional items in holding areas for future recycling or discard in more remote locations. The interior of the structural core is often demarcated by a rough stone wall or raised platform.

The structural core, including the living space inside and just outside buildings, is the most intensively maintained area of the house lot. Debris management within the structural core includes the sweeping of refuse out of the roofed-over area at least once a day. Large items of debris are placed in the interstitial zone mentioned above or dumped well outside this area. The structural core, an area demarcated by walls, platform structures, and zones of provisional discard, is the usual spatial focus of archaeological excavation.

The Clear Area

One of the areas outside the structural core, less commonly the focus of archaeological investigations, is the *clear area*, an extramural area of bare, hard-packed earth surrounding the house lot's dwelling structures. The clear area generally takes the form of a large kidney-shaped portion of the residential lot radiating out of the principal entryway into the structural core. It is highly significant that the size of the clear area in the sample of house lots from the Tuxtlas covaries with the activity regime of the household and not with the number of persons inhabiting the lot, the size of the dwellings, or even the overall size of the house lot (Killion 1987a). The clear area is referred to as the *patio* among peasant families in the Tuxtlas region and is considered an essential element of the proper *solar* or house lot.

Clear-area activities fall into two broad categories conducted within *staging* and *diversified activity zones* respectively. Staging zone activities include tasks associated with activities either initiated or completed outside the house lot. These activities include the preparation of farming implements for transport and use, arranging packs and loads for transport, temporary storage of items to be transported to distant locations, and other activities that produce little production debris. Rigging and arranging packs and quick tool repairs require clear open space. The staging zone is also the scene of celebrations, religious activities, children's play, and other leisure and recreational activities that overflow from the structural core into the clear area. These activities have public and social components that are best displayed from the staging area located close to the house lot's main entrance and exit. Heavy daily household use and maintenance of this zone lead to a clean and compacted soil surface in this portion of the clear area.

Within the clear area, the diversified activity zone is spatially segregated from the staging zone. This zone occupies a portion of the lot utilized for small-scale craft activities, food processing, socializing, and leisure. Activities conducted within the diversified activity area are shielded from public view (usually behind house-lot vegetation) in a portion of the lot away from access routes. It is swept and maintained in much the same way as the staging zone but not at the same intensity. Activities conducted within this zone include many small-scale projects of variable duration that are completed within the lot, including carpentry, protracted tool repair, fishnet mending and construction, boat building, and ceramic manufacture. Consequently, this area is maintained less rigorously and is swept in a patchy fashion. Drying and shelling of maize and beans take place within the diversified activity zone, which is generally located adjacent to the cook house. More debris is generated by the productive tasks conducted in this zone than in the staging zone. Sweeping removes the bulk of this material, but small pieces of debris become embedded in the surface of the zone, leaving a subtle record of clear area activities (cf. Hull 1987; Santley, this volume).

Generally, the diversified activity and staging zones do not overlap. Like the cooking and sleeping portions of the structural core, these two portions of the clear area are spatially segregated. Each zone is more or less emphasized in the structure of the lot, depending on the range of activities in which individual households engage.

The Intermediate Area

Production debris and household refuse generated by activities conducted in the structural core or clear area are managed by removal on a regular basis. Sweeping and refuse dumping result in the formation of both concentrated and dispersed midden deposits along the periphery of the clear area: the *intermediate area*. This portion of the house lot is characterized by the buildup of refuse swept and otherwise cleared from the structural core and the surrounding extramural clear area. The intermediate area forms a spatially extensive, sometimes discontinuous, ring of midden around the edge of the clear area. Within the intermediate area are several spatially restricted zones where refuse builds up into concentrated piles. Significantly, the size of the intermediate area varies in a complex fashion with the size of the other spatial components of the lot and with conditions outside the lot (Killion 1987a). Deal (1983) and Hayden and Cannon (1982) refer to a similar feature of Tzeltal site structure as the *toft*.

The intermediate area debris that accumulates around the edge of the clear area is a diverse collection of materials discarded within the lot. It is especially rich in sediment and hard durable refuse such as stone and broken ceramics. Softer refuse, such as food remains, either are consumed by dooryard animals or rapidly decompose due to humidity and heat. Other durable items forming the intermediate area assemblage include metal, glass, ash, charcoal, wood, ground stone, leaves, twigs, food, and fiber. Small fires are continually set within the more concentrated refuse piles of the intermediate area, reducing the volume of organic debris and generating more charcoal in the assemblage.

The Garden Area

Another portion of the lot where debris accumulates is the *garden area*. Gardens within house lots in the Tuxtlas region contain a polycultural mix of cultivated and wild species serving food, condiment, medicinal, construction, fuel, and ornamental uses. Generally, staple crops are not grown in dense stands within the house lot, although vegetable plots containing a few staples as well as nursery beds for field production are common. Garden areas also contain a large number of fruit trees. Fruit trees, bushes, vines, and herbaceous species form a dense net of vegetation around the peripheral portion of the lot, shielding the interior from view and providing a barrier to entry. Weeds grow in between the managed garden species and act to trap and stabilize debris that erodes off elevated portions of the intermediate area middens. Large items of debris are transported out of the core and clear areas and are provisionally or permanently discarded within the garden as well.

The dispersed "sheet" midden accumulation and vegetation that define the garden area account for a mean of 60 percent of house-lot space in the Tuxtlas study sample. In larger settlements, where house-lot space is more constrained due to crowding, household members also dump refuse generated within the residential lot at locations outside the lot. Common locations for dumping outside the lot include stream channels and vacant lots. Another common response to crowded conditions within more nucleated settlements is to consolidate house-lot

Table 6-1. Average Percentage of House-Lot Area Occupied by Use Areas in Portions of the Tuxtlas Sample

Sample Portion	House-Lot Component				
	Structural Core	Clear Area	Intermediate Area	Concentrated Refuse/Midden	Garden Area
South N=21	3.3	9.3	20.1	1.6	65.7
North N=19	4.3	17.8	16.5	2.7	58.7
Total N=40	3.7	14.6	19.3	2.2	60.2

refuse into concentrated piles within the intermediate area. Space-saving efforts such as these have a measurable effect on the size of other use areas within the lot, a topic to be treated later.

Table 6-1 summarizes the mean proportion of space devoted to each component of the house-lot model based on data collected in the house-lot survey of the Tuxtlas and stratified by two geographical regions. Following the garden area, the intermediate area is generally the second largest use area within residential lots in the Tuxtlas sample. The clear area represents the next largest use area within the lot, greatly exceeding the structural core and concentrated middens that represent just under 6 percent of residential space. Elsewhere it has been shown that patterns of refuse management observed at the surface of house lots correspond closely with the distribution of debris in sediments below the surface of the lot (Killion 1985, 1987a). The total weight of both surface and subsurface debris increases markedly in the peripheral portions of house lots outside the clear area and core (Killion 1987a).

Together the garden and intermediate areas represent almost 80 percent of the total amount of space within the house lots studied. It is surprising that areas *outside* the structural core have not received more attention in archaeological reconstructions of ancient settlement. In addition to being spatially dominant, these extramural and peripheral portions of the residential locus dominate the house-lot debris assemblage by weight. Inasmuch as intermediate and garden areas occupy a major proportion of individual house lots, it is reasonable to assume that these kinds of use areas account for a large proportion of the total space within entire settlements. Maintained public space, community access routes, and other kinds of interresidential space will alter this relationship. A great deal of interresidential space in contemporary Tuxteco settlements, however, consists of abandoned lots where vegetation and debris accumulate, more or less conforming to the overall settlement pattern being proposed. In addition, the wide avenues of contemporary communities are largely a product of Spanish colonial reorganization and not necessarily a feature of ancient settlement.

In general it is expected that intermediate and garden areas were the most extensive types of space within ancient settlements and generated a distinctive and spatially extensive debris signature. The extent to which variation exists with respect to the proposed pattern is an important indicator of a shift in community organization away from simple households organized for subsistence production to more diversified forms of production and more complex community structure. Before examining these propositions further with the data base from prehistoric Matacapan, the study will explore several factors influencing variability in the size of house-lot use areas in the Tuxtlas region today.

HOUSE-LOT USE AREAS AND HOUSEHOLD AGRICULTURAL ORGANIZATION

Diversity in the organization of local agricultural production within the Tuxtlas region provides an opportunity to test the strength and direction of the relationship between garden-residence site structure and the organization of agricultural production on nearby and more distant agricultural holdings. Variation in the amount of space devoted to different use areas within the lot can be seen in the southern and central geographical portions of the sample presented in table 6-1. The size of the structural core is somewhat greater in the northern portion of the region than it is to the south. Clear areas and concentrated middens are proportionally larger in northern area house lots while intermediate areas and gardens are generally larger in the southern house lots.

archaeologically because they represent such physically distinctive debris contexts. They are the areas within the lot where the lowest and highest amounts of debris accumulation are found and therefore represent the most clearly distinguishable elements of house-lot site structure. In the Tuxtlas the clear and intermediate areas vary greatly in size. The size of both intermediate and clear areas in the sample was found to covary most strongly with the intensity of cultivation on the household's off-residence agricultural plots.

Peasant households in the Tuxtlas engage in a variety of agricultural practices on plots situated close to settlements (infields) as well as at more distant (outfield) locations. Ethnoarchaeological research conducted in the Tuxtlas (Killion 1985, 1986, 1987a, 1987b, 1990) suggests that households occupying lots with small clear areas derive most of their subsistence from outfields—fields located greater than forty minutes' travel time from the residence. Large clear areas more typically associate with households emphasizing intensive infield agriculture found in the northern portion of the region.

Figure 6-3 depicts the spatial structure of a typical house lot from the northern portion of the region. Such households emphasize infield production and utilize extramural areas in conjunction with many activities going on at nearby fields. Hence, more clear space is necessary for an integrated domestic/agricultural regime. Figure 6-4 shows a more complex house lot from the southern portion of the Tuxtlas where clear areas are typically more reduced in size. When farmers focus their productive efforts at more outlying locations, as is customary in the south, many of the activities normally conducted at the house lot are shifted to outfields, where field huts are located, and less clear space is maintained for activities around the primary residence.

Table 6-2 illustrates the close relationship between the size of house-lot use areas and the intensity of cultivation at infield and outfield locations. Cultivation intensity is defined as the ratio of cropping to fallow time on each parcel. As suggested by the moderately high correlation coefficients of table 6-2, the level of cultivation intensity on infield and outfield parcels has important implications for the organization of household labor and residence in the Tuxtlas region. In fact, by calculating r-squared values for these correlation coefficients it becomes clear that anywhere between one-third to one-half of the variation in

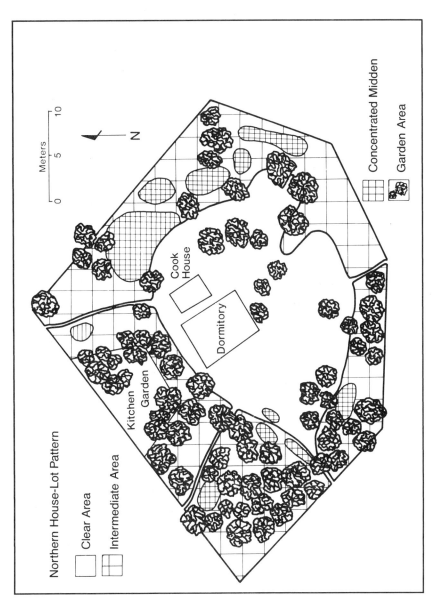

Figure 6-3. Simplified Plan of a Contemporary House Lot in the Northern Portion of the Tuxtlas Region

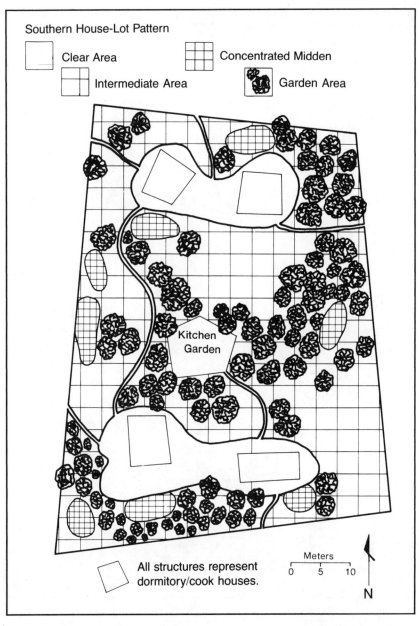

Figure 6-4. Simplified Plan of a Contemporary House Lot in the Southern Portion of the Tuxtlas Region

Table 6-2. Correlation Matrix (Pearson's r) of Site Structural Components and Factors of Agricultural Production

	Size of Clear Area (in sq. m.)	Size of Intermediate Area (in sq. m.)
Intensity of cultivation on infields (c/cf)[a]	.73	−.69
Intensity of cultivation on outfields (c/cf)[a]	−.59	.67
Time traveled to outfields (in minutes)	−.65	—
Size of concentrated middens (in sq. m.)	—	−.78

[a]ratio of years cropped to total number of years in one cropping-and-fallow cycle on a particular parcel of land

the size of house-lot use areas can be explained by variation in the intensity of cultivation at nearby and more distant field locations. Exploratory correlations such as these identify a subset of variables that show relatively more robust patterns of association among the total set of variables examined. Multiple correlation (addition of one or two more dependent variables to the correlation equation) greatly enhances the level of explained variation among variables in the Tuxtlas sample, yielding r values in excess of 0.8. Such analyses, presented in more detail elsewhere (Killion 1987a, 1987b, 1990), suggest a more complex level of interaction among the variables under examination. The correlations presented here identify basic trends in the data and support the preliminary examination of the variables presented below. As mentioned above, the size of the clear area increases when fields adjacent to settlements are intensively cultivated and decreases when farmers intensify production on more distant outfields. As might be expected, clear areas also decrease in size as farmers travel greater distances to outfield locations.

Table 6-2 also indicates that the intermediate/midden area is sensitive to the intensity of cultivation on outlying and nearby agricultural plots. Intermediate areas systematically expand and contract in an inverse pattern compared to the clear area; that is, the dispersed pattern of refuse disposal within the intermediate area produces a larger intermediate area. Larger intermediate areas are common in the southern portion of the Tuxtlas where settlements are surrounded by poor-quality agricultural lands that are less intensively cultivated. As a consequence, less pressure occurs on the use of space in the settlement-infield zone. Village settlement is more dispersed and the use of residential space is more relaxed.

In the northern portion of the study area, settlements are larger and more nucleated. Residential space within northern area towns is restricted due to crowding caused by more intensive use of flat, highly valued infields. These conditions favor more space saving and efficient debris management within residential areas because the overall amount of space available for settlement is smaller and population density within the town is high. Not surprisingly space-saving piles of concentrated refuse expand as the overall size of the intermediate area shrinks, thus resulting in the negative correlation between intermediate area and concentrated refuse middens in table 6-2. Data on overall house-lot size and community nucleation suggest that households tend to concentrate refuse in middens rather than in spatially extensive intermediate areas as the overall size of the house lot decreases.

The patterns described above suggest a structural reorganization of residential space in response to agricultural conditions outside the residential area. In reality, a host of environmental and economic factors interact with the organization of household economy in the Tuxtlas to produce a distinctive set of local agricultural systems based on infield-outfield production (Killion 1987a, 1987b, 1990).

Regular patterns in the relationship between the use of space within residential areas and in agricultural zones surrounding settlements provide some provocative information concerning the relationship of behavior, the organization of agricultural production, and the form or layout of residential settlement. The contrast between refuse-laden and refuse-free space is an operational measure of the normally "invisible" factors of agricultural production so important to identifying the structure of ancient agricultural systems.

The distribution of household debris in the house-lot model can be conceptualized as a series of concentric rings extending outward from the residence. Each ring farther out from the core contains more household-produced and -managed debris until a threshold is reached in the intermediate area or garden, after which the pattern can be repeated in another residential lot or the density of material falls off in crop land adjacent to residential zones. In the section to follow, the implications of variation in this pattern of residential site structure as well as the spatial dominance of extramural areas outside the structural core are examined in light of the archaeological record of settlement at ancient Matacapan.

AN APPLICATION OF THE HOUSE-LOT MODEL TO THE SURVEY DATA FROM MATACAPAN

The archaeological site of Matacapan is located in the central flatlands (see fig. 6-1) of the Tuxtlas between the modern centers of San Andres and Catemaco. Prehistoric Matacapan was divided by the deep channel of the Rio Grande de Catemaco, which flows out of a large lake some five kilometers east of the site. Weathered volcanic soils, derived from the surrounding uplands, accumulated in this area throughout the Holocene, creating one of the largest expanses of flat arable land anywhere in the Tuxtlas. These conditions, along with high yearly rainfall, combine to form a rich agricultural resource utilized by both modern and prehistoric populations occupying the central portion of the region.

Figure 6-5 shows a portion of the overall site map centered on the monumental core of Classic period Matacapan. More than 100 prehistoric architectural mounds are found on the intensively cultivated agricultural fields surrounding the modern village of Matacapan. Intensive surface survey by members of the Matacapan Archaeological Project has revealed more than twelve square kilometers of residential settlement surrounding the core, as evidenced by low residential mounds and a continuous scatter of ceramic and lithic debris. A pro-

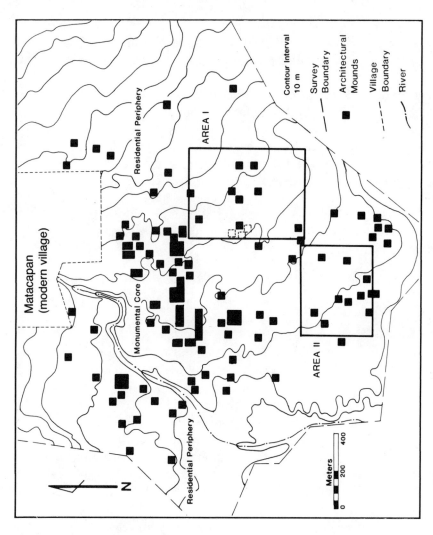

Figure 6-5. Topography and Mounded Architecture within the Matacapan Archaeological Zone. Study Areas I and II are indicated.

gram of surface collections consisting of three-by-three-meter units positioned ten meters apart along transects spaced approximately every seventy-five meters has been implemented across the entire twelve square kilometers.

Given the wide spacing of these transects (probably greater than the size of most cultural units under consideration), it is expected that the survey will systematically underrepresent smaller components of residential site structure (the structural core, concentrated middens, and possibly clear areas) and overrepresent larger elements of site structure (gardens and intermediate areas). This problem has recently been discussed by Stark (1987), who chose to space survey personnel eighteen meters apart in the Mixtequilla, Veracruz, survey in order to pick up cultural entities down to twenty meters in size. Historic plowing and more than a millennium of continuous agricultural utilization at Matacapan act further to complicate site structural patterns and require careful consideration by the archaeologist.

Settlement Structure at Matacapan

The results of the intensive survey at Matacapan are graphically presented in figures 6-6 and 6-7, which show spatial patterns in the density of all Classic period ceramics in two smaller portions of the site: south (Area I) and east (Area II) of the monumental core. Figures 6-6 and 6-7 are isopleth maps of surface ceramic densities for Areas I and II. The contours in these plots represent the borders of areas corresponding to portions of the frequency distribution of sherds across the surface of the site divided into quartiles. The total number of sherds per surface collection in the survey (N = 6,500 units) ranged from 0 to 1,875. The mean number of sherds per square meter (29) reflects the inflationary effect of a relatively smaller number of collection units with very high sherd counts. The sample median (17 sherds per square meter) is a more accurate representation of the average number of sherds standardized across the surface of the site. Quartiles break the distribution down into two equal-sized ranges on either side of the median, yielding the four zones indicated in figures 6-6 and 6-7. Mounded architectural features have also been plotted on the maps along with the distribution of ceramic debris.

Figures 6-6 and 6-7 illustrate a number of site structural patterns corresponding to expectations of the house-lot model. Generally, mounded architecture falls within or adjacent to areas of concentrated debris, or the first quartile of the ceramic density distribution (A1–3 and B1 in fig. 6-6, A1–3 and A7–8 in fig. 6-7). Concentrated midden areas, such as these, are commonly associated with areas that are relatively clear of debris, corresponding to the fourth quartile of the distribution (e.g., A2, B1 in fig. 6-6, A1–3 in fig. 6-7). The spatial association of high- and low-density portions of the ceramic distribution conform closely to the spatial association of structures, clear areas, and midden/vegetated areas stipulated in the house-lot model and observed in the Tuxtlas house-lot sample.

Residential complex B1 in figure 6-6 resembles the ideal house-lot model. The mound, probably representing a building platform that supported residential structures, separates the spatially segregated midden and clear areas. Midden areas (first quartile) in figure 6-6 are generally larger than the midden areas represented by the first quartile in figure 6-7. Surrounding the midden, clear area, and structural core of complex B1 is a low-density zone of ceramic debris (fewer than seventeen sherds per square meter) enclosed within the contour of the second quartile of the distribution. The spatial position of the second quartile (surrounding mounded architecture, clear areas, and concentrated midden deposits) suggests that this area corresponds with the intermediate area discussed earlier. Furthermore, the second quartile of the ceramic distribution is absolutely greater in size than the first and fourth quartiles and also appears greater in size with respect to the size of the first quartile (midden/concentrated refuse) in Area II (fig. 6-7) than in Area I (fig. 6-6).

The third quartile of the ceramic debris distribution is spatially the most extensive zone in both Areas I and II. Second and third quartiles combined clearly represent the most extensive debris patterns at the site. A spatially extensive scatter of low-density ceramics is exactly the situation expected if the distribution of garden and intermediate areas spatially dominated prehistoric settlements as they do contemporary settlement space. As already mentioned, smaller components of site

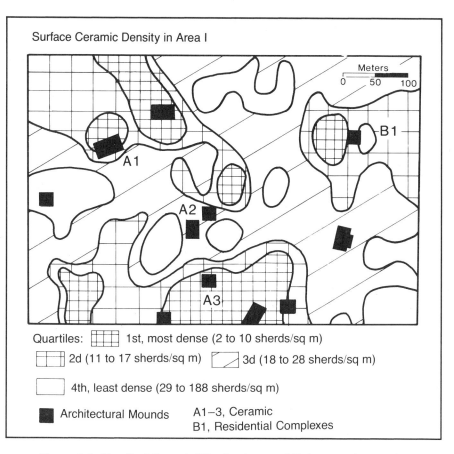

Figure 6-6. Detail of Ceramic Distributions and Habitational Mounds in Study Area I

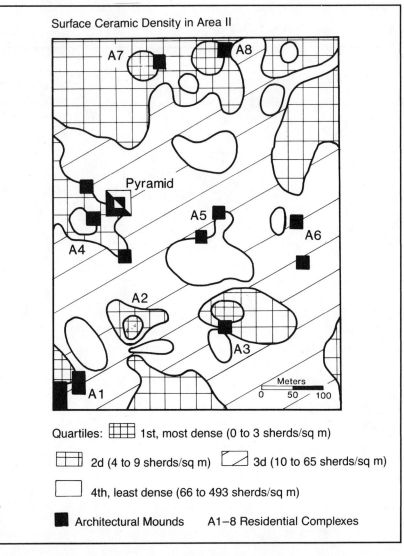

Figure 6-7. Detail of Ceramic Distributions and Habitational Mounds in Study Area II

structure, such as clear areas and concentrated middens (first and fourth quartiles), will be underrepresented because of transect spacing. Likewise, the structural core (platforms) will also be underrepresented either because architectural mounds were destroyed through plowing or because the structures were not visible on the surface. Nonetheless, both of the smaller quartiles combined (first and fourth) do not approach the spatial extent of the second and third quartiles of the distribution in Areas I and II, so that the ceramic debris patterns generally conform to the expectations of the house-lot model for settlement structure.

The Use of Settlement Space at Matacapan

Surface distributions of prehistoric ceramics at Matacapan are the result of multiple processes. Behavioral patterns related to the use of residential areas have been outlined above. Initial discard and deposition of ceramic debris have been related to the maintenance of clear activity space and the management of refuse areas within residential lots. Equally important to the interpretation of survey data from Matacapan is the assessment of taphonomic processes and data recovery biases.

A number of basic assumptions related to large-site taphonomic processes and analytical biases due to large-scale sampling guide the discussion that follows. Taphonomic processes affecting the interpretation of surface data include modern agricultural and construction activities (plowing and land leveling) that: (1) decrease the number of prehistoric residential buildings and other structures recorded under the category of mounded architecture, (2) decrease the size of maintained clear areas by spreading material over the surface, and (3) increase the size of managed midden areas, also due to the spread of materials across the surface of the site.

Data recovery biases related to transect spacing result in: (1) the overrepresentation of larger spatial entities, such as the second and third quartiles of the ceramic distribution, and (2) the underrepresentation of smaller spatial entities, such as the first and fourth quartiles of the ceramic distribution. Any consideration of proportional differences in the spatial distribution of surface ceramics at Matacapan must take these biases into consideration.

Table 6-3 lists the spatial proportions of the surface ceramic distribution, broken down into quartiles, and the proportions of space stipulated by the house-lot model from the contemporary house-lot survey. Use areas and quartiles are arranged with least debris or least number of sherds on the left, increasing to the areas containing the densest amount of debris or sherds on the right.

Mounded architecture and the structural core of the house lot occupy relatively small proportions of settlement or residential space and in fact are roughly equivalent in table 6-3. Given the destruction of pyramids and building platforms, which are difficult to miss on survey, it is quite likely that the "built" portion of the site was probably greater in Classic times. Taking this factor into consideration would increase the original number of mounds at the site and argue for a greater number of residential structures in both Areas I and II, with somewhat thicker settlement in Area II.

The clear area/first quartile comparison suggests that the values for debris-free space across the site and within the house-lot model are within the same order of magnitude. It is expected that clear, maintained settlement space will decrease in spatial extent over time, due to debris spread, and will be underrepresented as an artifact of sampling. Actual values for clear areas were therefore greater than those presented in table 6-3. Area II clear areas might have actually approached the values reported for the modern house lots. Area I clear areas would have probably exceeded the average size of clear areas found in modern settlements.

The first and fourth quartile distributions indicate that Area II was more densely settled than Area I with relatively less space maintained free of debris. Relatively less built space occurred in Area I; however, clear areas covered proportionally more space in Area I than in Area II. This pattern contrasts sharply with contemporary experience in the Tuxtlas where denser, more nucleated settlements have larger clear areas. Where settlement is less nucleated today, clear areas are proportionally smaller. The modern patterns appear to be precisely the opposite of the archaeological ones. This contradiction suggests that the

Table 6-3. Average Areal Percentage Comparison of House-Lot Model Components and Prehistoric Ceramic Density Distribution Quartiles

	Unit of Analysis	Clear Area	Garden Area	Intermediate Area	Total of Garden and Intermediate	Concentrated Refuse Area	Structural Core
Contemporary data	House lot	14.6 (least debris)	60.2	19.3	79.5	2.2 (most debris)	3.7
		1st Quartile	2d Quartile	3d Quartile	Total of 2d & 3d	4th Quartile	Mounded Architecture
Prehistoric data	Area I	14.0	47.2	27.5	74.7	10.7	2.5
	Area II	9.4 (least sherds)	58.0	25.9	83.9	2.8 (most sherds)	3.7

house-lot model proportions are not the appropriate comparative base line for *all* segments of the Matacapan community. The following discussion of the more debris-laden spaces at the site helps to explain this apparent contradiction and the variation in clear space across the surface of Matacapan.

The distribution of dispersed ceramic debris at the site conforms more closely to refuse patterns established in the house-lot model. Much like garden areas in contemporary house lots, the area covered by the second quartile distribution at Matacapan accounts for more site space than all other quartiles listed in table 6-3. The third quartile is the next largest component of site space, corresponding to the intermediate area in the house-lot model. Because the dividing line between the second and third quartiles is essentially arbitrary, these two spatial divisions of the ceramic distribution will be summed for the purposes of this discussion. When combined, the total area of dispersed ceramic debris across the site is proportionally equal in size to the combined size of the garden and intermediate areas of modern house lots. Like the pattern observed in the northern and southern portions of the house-lot sample, the larger area of dispersed ceramic debris at Matacapan is associated with relatively smaller clear areas while a reduction in the total area of dispersed refuse corresponds with a proportionally larger area of clear space. This pattern indicates that the management of debris by residential groups occupying Area II was more like the management of debris within contemporary subsistence-oriented agricultural households than the pattern of debris management in Area I.

The most concentrated accumulations of ceramic debris at the site also suggest variation in the way debris was managed within Areas I and II at Matacapan. Concentrated refuse areas are an areally small component of contemporary house-lot site structure, and it is expected that similar elements of prehistoric site structure would be systematically underrepresented in the archaeological sample. This sampling problem is offset, in effect, by taphonomic processes that tend to increase the size of the fourth quartile over time, making the proportions of concentrated refuse and the fourth quartile roughly comparable. The size of the fourth quartile in Area II is of the same order of magnitude observed in the subsistence-oriented house lots of the contemporary sample. The amount of fourth quartile space in Area I, however, is much greater than that observed in Area II or by the concentrated middens in contemporary settings.

A knowledge of the technological attributes of the ceramic assemblage in Area I explains the larger size of both the first and the fourth quartiles and suggests basic economic and productive variability within the prehistoric community of Matacapan. It appears very likely that the residential entities that made up Area I at Matacapan were specializing in large-scale ceramic manufacture. The presence of a large number of kilns, ceramic manufacturing errors, and a limited number of highly standardized ceramic types support this proposition (Santley, Ortiz, and Pool 1987). Workshop groups occupying this portion of ancient Matacapan would have regularly produced large amounts of ceramic debris, explaining the middens adjacent to the building platforms (A1 and A3) in figure 6-6. Postdepositional taphonomic dispersal of these middens would obliterate most of the clear maintained space adjacent to residential/workshop complexes, although inspection of Area I shows that all A-group mounded architecture is within fifty meters of either third quartile or fourth quartile low-density scatters. The management of ceramic production space within the occupational units of the A group would also require more clear space for ongoing ceramic manufacture. The need for clear space within production areas has already been indicated on a smaller scale by the presence of the diversified activity zone of the house-lot model within house lots in the Tuxtlas today. The larger proportion of fourth quartile clear space within Area I is posited here as the diversified activity zone writ large. Because the sum of all dispersed debris (second and third quartiles) is reduced in Area I in comparison to Area II, it is expected that the use of Area I for garden space within the settlement was also reduced.

Among the residential units of Area II, the spatial expansion of dispersed refuse probably supported a higher density of vegetated space, much of it managed horticultural space. In Area II clear areas are reduced, dispersed refuse areas expand in size, and no obvious evidence exists of occupational specialization. It is posited that the residential groups of Area II were made up of households oriented toward subsistence production and that the low-density debris scatter that spatially dominates Area II is the archaeological consequence of one component of that subsistence system.

CONCLUSIONS

Ethnoarchaeological research in the Tuxtlas has shown that the regime of agricultural production, especially the locational focus of production in peasant households, affects the way space is utilized at the residence. This relationship is especially visible in the way debris is managed at different locations around the house lot, providing the structural basis for the house-lot model. The model, in turn, appears to correspond closely with surface and subsurface debris distributions from house lots in the Tuxtlas region and can be used to understand variability in these patterns.

Based on the material consequences of the use of space, the house-lot model has important implications for the archaeological record. Variability in the distribution of surface debris at Matacapan has been related to possible differences in the use of space between two segments of the ancient community at the site. The pattern observed in Area I relates to a portion of the settlement where specialized ceramic-producing households predominated. These households left a distinctive pattern of ceramic debris related to crowded residential conditions in between production areas and workshop middens. In Area II farming households probably predominated. Residential spacing and disposal patterns were relaxed relative to Area I and the ratio of dispersed to concentrated refuse areas suggests a focus on infield cultivation. Below the vegetated cloak of the typical lowland "garden city," Classic period Matacapan had developed into a productively diversified urban community.

This preliminary reading of the archaeological record at Matacapan demonstrates that the house-lot model is a useful tool for examining internal variability in community structure at a Classic period regional center in the Tuxtlas region. Inferences have been made concerning settlement structure and agriculture as well as more specialized forms of production engaged in by the ancient Matacapeños. Future research efforts must be directed at broadening our understanding of the material consequences of the use of space in contemporary contexts, including both residential and agricultural space, before such inferences concerning prehistoric contexts can be more generally applied to the archaeological record.

CHAPTER 7 ROBERT S. SANTLEY

A Consideration of the Olmec Phenomenon in the Tuxtlas

EARLY FORMATIVE SETTLEMENT PATTERN,
LAND USE, AND REFUSE DISPOSAL AT
MATACAPAN, VERACRUZ, MEXICO

MANY SCHOLARS consider the southern Gulf Coast of Mexico as the heartland of one of the earliest "complex societies" in the New World. This complex society, often called the Olmecs in the literature, refers to a people who shortly after 1350 B.C. "built massive pyramids and other nonresidential structures, aggregated in relatively large settlements, engaged in a variety of specialized arts and crafts, and invested considerable energy and resources in drainage channels and terraces in order to intensify agricultural production" (Wenke 1984:352). Despite the enormous amount of research over the past fifty years on the southern Gulf Coast, little is known about the smaller sites that supported the major centers.

This chapter describes the evidence on Olmec settlement and land use during the Early Formative period at Matacapan, an archaeological site located in the Sierra de los Tuxtlas of southern Veracruz, Mexico (fig. 7-1). Data from Matacapan indicate that Olmec land use involved both house-lot gardens (or garden-orchards) and infields situated at various distances from habitation sites, producing a dispersed settlement pattern. This information comes from two sources. Analysis of materials from surface assemblages indicates the presence of at least ten habitation sites as well as Olmec remains in off-site contexts. Excavations in a number of areas also exposed Olmec "living surfaces."

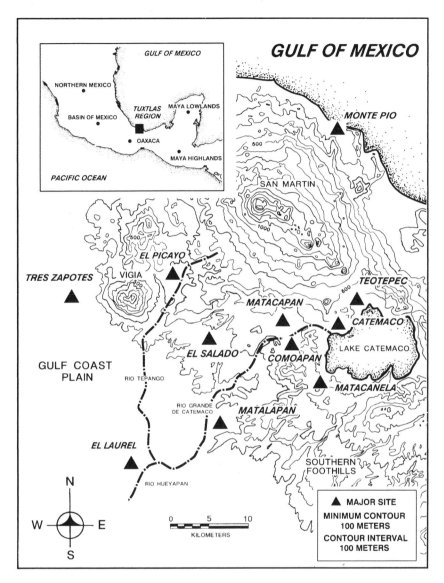

Figure 7-1. The Western Tuxtlas

These occupation surfaces include the remains of Formative period structures and bell-shaped pits, agricultural features, and associated trash middens, all sealed by a layer of volcanic ash.

My presentation of this evidence is organized in several parts. First, I discuss patterns of land use in tropical forest Mesoamerica. Next I describe the general pattern of Olmec settlement at Matacapan. While most of the Olmec material derives from residential sites, a substantial amount comes from off-site localities (i.e., from areas between habitation sites), indicating use of the landscape, probably for agricultural purposes. The third section discusses the Olmec living surface south of Mound 22 in the central portion of Matacapan and compares the archaeological patterning to various models of contemporary house-lot structure and refuse disposal. I conclude with a prospectus for future research.

CONTEMPORARY LAND USE IN TROPICAL FOREST MESOAMERICA

Agriculture and horticulture are land-use strategies in which a natural plant cover is replaced by one that is economically useful to humans. In the humid tropics, reliance on cultivated resources generally involves a number of land-use strategies. Nonurban settings frequently have a *garden* or a *garden-orchard* where vegetables, root crops, fruit trees, economically useful shrubs, condiments, medicinals, and sometimes cereals are grown. These plots may be distinguished from fields by their polycultural emphasis and by their usual proximity to residences. Depending on the amount of space available, prevailing tenure arrangements, and environmental factors, the garden or garden-orchard may have a multitiered structure, as is the case often in the Tuxtlas today (Killion 1987a), or it may be planted in a variety of annuals grown in different parts of the plot, a cropping arrangement more common in central Mexico (Sanders et al. 1970). Under conditions of low population density, the garden-orchard provides the bulk of all cultivated foodstuffs consumed by the household. Often these garden-orchards are located some distance from the residence, but, as population densities rise, gardens or garden-orchards tend to be situated on or near the house lot.

While the garden or garden-orchard may produce as much as one-third of the foodstuffs the household needs in high-density settings, the remainder must come from agricultural plots located outside the settlement (Sanders and Price 1968). These plots are differentially utilized, depending on soil fertility, local edaphic and hydrologic characteristics, rainfall and flooding patterns, temperature, population density, and other sources of household income. In an area like the Tuxtlas these plots fall into two classes (Killion 1987a). *Infields* are parcels situated within a forty- or forty-five minute walk or "travel time" from the farmer's residence, whereas *outfields* are usually located farther away. Infields are generally located outside settlements, though in some cases they occur within villages. Fields may be extensively cultivated, or they may involve intensive practices such as short fallowing, hydraulics, or raised fields. These plots are commonly planted in staples, frequently annuals such as cereals, which provide the bulk of the household's subsistence needs and caloric requirements. Although agricultural intensity may vary as a function of distance from the community, such patterning is not always the case with more intensively cultivated parcels situated farther from the settlement.

When population density is low, most cultivated resources are grown on a single plot that is shifted every few years, particularly in areas where no handy means of natural or artificial soil enrichment exists. As population density levels rise, however, prime lands are at a premium. In areas where zonal and topographic variability occurs within short distances of settlements, like the southern Tuxtlas, sites are commonly located in less productive zones. Here infields are cultivated less intensively because they are less susceptible to intensification without terracing or other erosional control devices and because the outfields, located on highly productive river levees, are more intensively exploited without much artificial enrichment, due to the annual deposition of alluvial sediments. In addition, because fewer constraints exist on the amount of space around communities, villages are more dispersed, and individual peasant house lots are comparatively large.

In parts of the Tuxtlas where local agrozonal variability is not as pronounced, for example in the flatlands and gently sloping terrain around Matacapan today, intensity of use appears to be conditioned mainly by population density and the amount of nonsubsistence use.

High population densities produce extreme competition for space, a condition exacerbated by the commercial cash cropping of tobacco by wealthy landowners in the area today. Infields around agricultural villages are consequently cultivated at higher levels of intensity relative to outfields, and because infields are intensively cultivated, more constraints occur on the amount of available space thus affecting a reduction in house-lot size and a comparatively more nucleated, village settlement pattern.

While the subsistence base of many village agriculturalists in the Tuxtlas involves an emphasis on gardening and infield-outfield agriculture, deriving the archaeological consequences of this system of land use is a much more difficult matter. Clearly, finding relict or fossilized agricultural features (e.g., raised fields, irrigation networks, or cultivated surfaces) will only occur under the most favorable of circumstances. However, as I will argue below, an approach that incorporates information on contemporary land use, refuse disposal, and house-lot structure, their material correlates, and variation in archaeological patterning across the landscape provides a better basis for deriving inferences about Olmec economic organization than traditional perspectives.

EARLY FORMATIVE SETTLEMENT PATTERNS AT MATACAPAN

Data on Olmec settlement patterns at Matacapan come principally from surface survey (Pool et al. 1986; Santley et al. 1984, 1985a, 1985b; Santley, Ortiz, and Pool 1987).[1] Although this occupation spans the Formative period, most Olmec sites discovered thus far are Early Formative in date (ca. 1400–1200 B.C.). The fact that Early Formative sites are relatively common suggests population growth over pre-Formative levels: an inference supported by recent palynological work that indicates progressive increases in the amount of *Zea* represented as well as decreases in the pollen counts for arboreal species, implying forest clearance (Steve Nelson, personal communication, 1988). This occupation was also socially differentiated, to judge from the presence of "status" ceramic wares, ball-game paraphernalia, and miniature basalt heads.

THE OLMEC PHENOMENON

The Early Formative occupation in the Tuxtlas was abruptly terminated by a massive explosion of the volcano, San Martin, that devastated the local population, covered much of the area with a deep layer of black ash, and preserved cultural features such as agricultural surfaces. The effects of this eruption appear to have been the most catastrophic in flatland (i.e., valley) locations that were not substantially reoccupied until Classic times. Late Formative sites seem to be restricted to slopeland situations where colluvial action, erosion, and weathering would have removed the ash layer fairly rapidly and where the rate of forest regrowth consequently would have been the greatest.

Early Formative material is less common in surface contexts at Matacapan than Classic period material. This relative lack of material is due in part to superpositioning by later components that often swamp the Formative assemblage, combined with the fact that pottery and obsidian are generally less common in Formative contexts. Hence, Olmec ceramics account for only about 0.42 percent of all ceramics from the intensive survey (i.e., 566 sherds) or 7.84 percent of all rimsherds. The density of Early Formative pottery averages $0.43 +/- 1.01$ sherds per square, but the density for collections with Olmec materials ranges from 1 to as many as 19 sherds (mean $= 2.30 +/- 1.36$ sherds). Samples with Olmec material are distributed throughout the area surveyed and occur either as isolated squares or in aggregations.

To describe this frequency distribution the following method was employed. All survey transects containing squares with Olmec material were ranked by density and placed in quartiles. A quartile is a monothetic grouping containing one-quarter of the list of units ranked in order from smallest to largest. All squares in the first two quartiles (>0–25 percent and 26–50 percent) were defined by the presence of a single sherd, and 75 percent of these occurred as isolated squares (i.e., no other squares in the transect contained Early Formative material). Densities for squares from transects in the next two quartiles, in contrast, were much greater: 1.50 and 3.58 sherds respectively. Moreover, isolated squares accounted for only 30 percent of these samples, and in many cases these isolated squares were located near transects with other squares containing Formative ceramics.

The distribution of squares with Olmec ceramics is illustrated in figure 7-2. As is readily apparent, squares with higher densities tend to

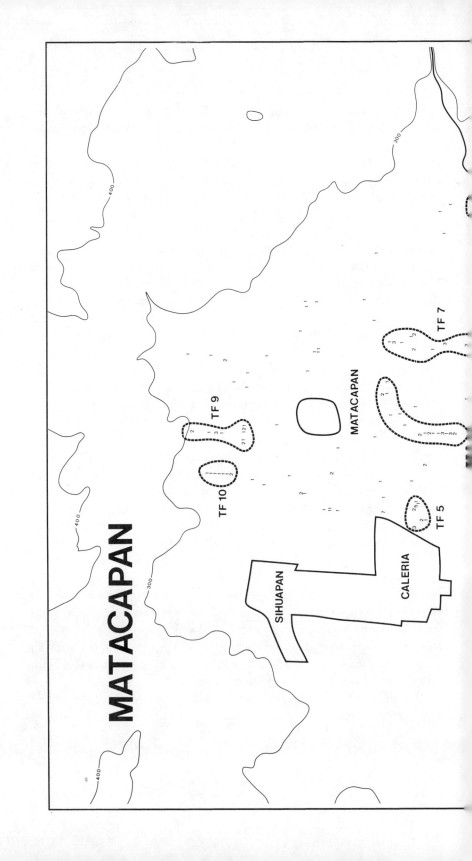

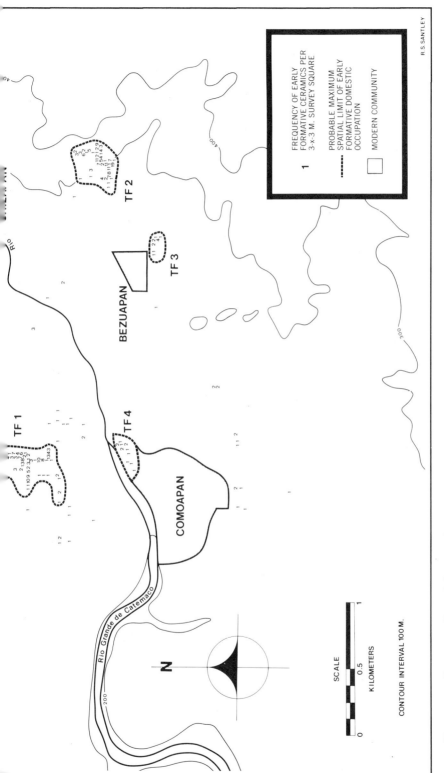

Figure 7-2. Formative Period Sites in the Matacapan Area

cluster, whereas low-density squares are distributed throughout the area surveyed. Altogether ten aggregations of squares exist, two (sites TF1 and TF2) are much larger in size and defined by much higher densities of material. The Olmec occupation levels at TF1 and TF2 are very close to the present-day surface: hence the increase in surface density. On the other hand, the Early Formative levels at other sites are more deeply buried, in some cases by as much as three meters of later deposits. Excavation has demonstrated that the densities of material at different Early Formative sites are very comparable. Although I lack excavation data for all sites, subsurface testing in and around several sites suggests that the boundaries depicted in figure 7-2 are close approximations to the true limits of the areas defined by dense distributions of refuse. In other words, locations where little or no Olmec pottery exists on the surface are also locations where almost no Formative material occurs under the ash layer, and vice versa.

Inspection of an Early Formative map indicates that settlement in the Formative period was more dispersed than in the Middle Classic when almost all of the area surveyed was covered by public buildings, house mounds, extramural activity areas, refuse dumps, and other cultural features. The Early Formative occupation occurs at ten localities that vary in size. As pointed out above, TF1 and TF2 are much larger than the others, and these sites produced almost all of the evidence suggestive of status differences. Although no Formative period public buildings or platform residences were discovered during the survey, decorated ceramics are more common at these localities. TF1 also produced several basalt ballcourt yokes that are Formative in date (Jeffery Wilkerson, personal communication, 1985), and a miniature basalt head similar to ones from San Lorenzo is said to come from this location. These two sites account for 64.8 percent of all Olmec ceramics retrieved. An additional 18.7 percent of the assemblage comes from the remaining eight sites, while another 16.4 percent derives from off-site locations: from areas beyond the boundaries of what would normally be considered residential sites. The relative amounts of Olmec material from other site and off-site contexts are probably even greater than the above estimates would suggest, given the fact that most of these samples come from areas of Matacapan where Early Formative levels are deeply buried. The total momentary population represented by all sites probably did not exceed 1,000 persons.²

Spatial analysis of the Formative materials produces additional patterning. Olmec sites at Matacapan are situated 697+/−230 meters apart. The nearest neighbor statistic for the central points of all aggregations of squares is 1.24, indicating a tendency toward regular spacing. Systematic site distributions of this sort are expected under conditions of mutual repulsion and competition between units for space (Earle 1976). Such conditions might be expected if the areas around sites were utilized rather intensively and areas farther away more extensively. Newly founded sites would be expected to be located in extensively utilized zones, and, with additional demographic growth, intensification of use directly around them would serve to "repulse" new settlements outward. On the other hand, the nearest neighbor statistic for all off-site squares is 0.62, indicating a marked clustering tendency. The mean observed distance between off-site squares is 130+/−129 meters, whereas the expected mean given the array of points is 210 meters. The mean distance between squares exhibits little change from north to south, although the array in the south appears to be more dispersed (north = 126 +/− 87 meters; south = 135+/−172 meters). This difference, however, is not significant at any commonly accepted confidence interval ($t = -0.2873; p > .2$). Clustered patterning of this type is expected in situations where materials are discarded in a context of attraction to resource zones, situations that might occur if the prevailing system of agriculture emphasized the regular reuse of off-site areas.

The average distance from off-site squares to the nearest site boundary is 415 +/−259 meters. A histogram plotting the distances of all squares, however, shows that the distribution is really bimodal, indicating two distinct spatial zones of material accumulation (fig. 7-3). Ceramics are most often discarded at 101–300 and 501–800 meters from sites than in any other catchment. Although interim squares also contain material, these two modal ranges account for 80 percent of all off-site samples. This bimodality probably reflects the presence of an agricultural system that involved the use of two different components of the landscape. Off-site locations where material was deposited near

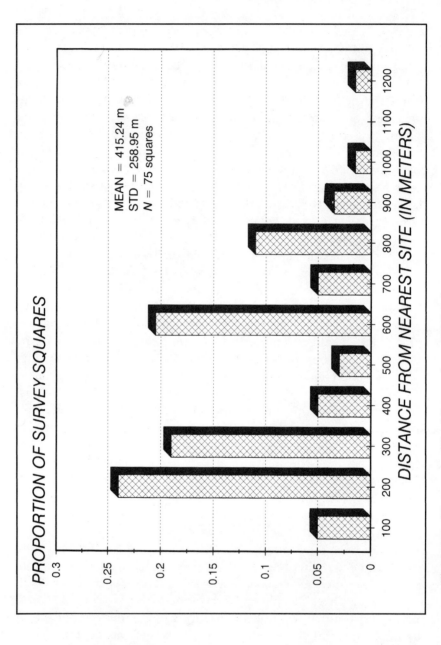

Figure 7-3. Histogram of Total Pottery from Off-Site Squares

sites may be conceptualized as "inlying infields," whereas more distant areas were probably "outlying infields," inasmuch as the distance to these locations from sites is not particularly great. Outfields may have also been present.

The evidence at hand consequently suggests that an "infield" agricultural system was probably in use during the Early Formative period. Infield plots are also frequently monocultural (i.e., planted in maize), though secondary cultigens such as beans and squash or occasionally economically useful trees and shrubs may be intercropped. Such a system would be expected under conditions of moderately high population density but where the majority of each household's livelihood still came from agriculture, not from occupational specializations such as wage labor or craft production. Although I have yet to devise methods effectively to monitor occupational roles such as wage laborer, it does appear that craft specialization was not a major economic role at Early Formative Matacapan.

Patterning is also evident when the samples are stratified by spatial zone. In the northern half of the area surveyed the mean distance to off-site squares is $378 +/- 192$ meters. The distribution of squares is still bimodal, with peaks at 101–200 and 501–600 meters (fig. 7-4a). In the south, off-site squares tend to be located farther away from the boundaries of habitation sites (mean = $466 +/- 323$ meters). This distribution, however, is even more bimodal; peaks occur at 301–400 and 701–800 meters, with no cases in the 401–700 catchment range (fig. 7-4b). The difference in the degree of modality in the distributions (i.e., the degree of spread between the modes) may be related to site density. The northern half of Matacapan contains the greatest number of sites, and expectably the average distance between nearest neighbor sites (mean = 619 meters) is less than in the south (mean = 815 meters). Productivity on Matacapan's flatlands, which tend to be situated in the north, is also somewhat higher than on the slopelands in the south. Flatlands in the north are also the areas favored today for high-grade tobacco, a very nutrient-demanding crop when planted on an annual basis (Miguel Turrent, personal communication, 1985). Hence, all other things equal, it would appear that areas around sites in the north could be cultivated at somewhat higher levels of intensity than in the south, a situation that might affect a "pulling" in of areas utilized and thus a decrease in the distance between habitation sites.

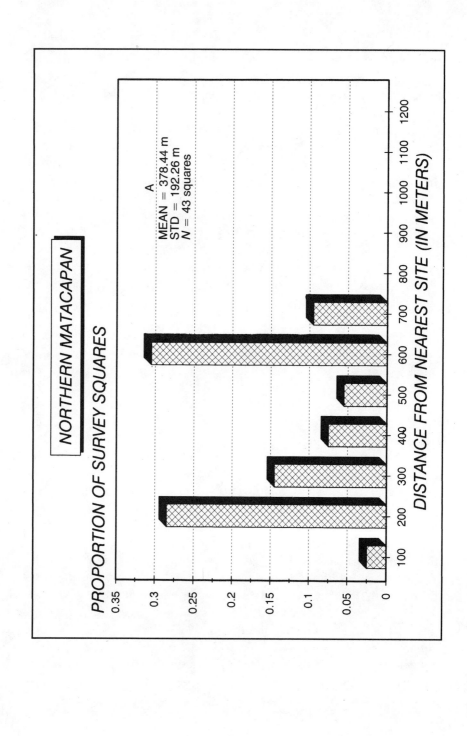

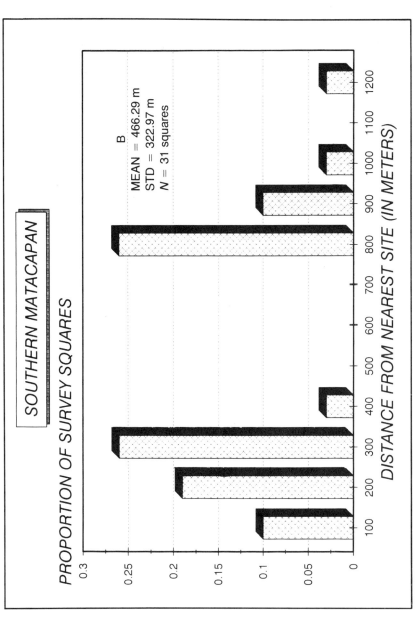

Figure 7-4. Histogram of Total Pottery from Off-Site Squares Stratified by Spatial Zone

How do these distributions compare with patterning on infield use in the Sierra de los Tuxtlas today? Contemporary land use in the Tuxtlas emphasizes infield-outfield agriculture, with garden-orchards located on house lots in rural villages (Killion 1987a). Distances from house lots to agricultural plots, however, were measured in travel time, not distance, by Killion because modes of transport vary. While farmers will often use bus transportation or horses to travel to outfields, transit to infields, especially those relatively near the community, is generally by foot. To transform travel time to distance, a walking rate on level ground of 2.5 miles or 4.02 kilometers per hour was assumed. Although infields today may be situated as distant as 2.68 kilometers from the residence, the vast majority (82.5 percent) occur within 1,400 meters. Of these, 82.4 percent are located within 700 meters of the community (i.e., up to a ten-minute walk). Assuming that off-site squares indicate the use of off-site agricultural plots, 75.8 percent of the archaeological outliers fall between >0 and 700 meters from the boundaries of habitation sites. Though the sample sizes are small ($N_1 = 33$ and $N_2 = 74$ respectively), the difference of proportions test yielded insignificant results ($Z = -0.7937; p > .4$), implying similar patterns of utilization of off-site localities near communities in the two samples. The difference of means test produced similar results ($t = 1.2416; p > .2$), as did the chi-square test ($X^2 = 0.62; p > .3$; phi $= 0.0761$).

The assemblages from site versus off-site contexts also vary in internal composition. All samples contain the same array of service and utility wares, but they differ in terms of the proportion of each represented. Part of this variation may be microchronological, but it must be acknowledged that the number of specimens from some contexts at Matacapan is not particularly large (total $N = 566$), so small sample size may be a problem when the collection is broken into a large number of ceramic wares. Grouping the wares into classes that reflect the wares' primary function, food consumption (plates and bowls) versus food preparation (*tecomates*, or spherical neckless jars) and short-term storage (jars), is one way around this problem. Vessels used for food consumption are common at habitation sites and in off-site contexts. *Tecomates* and jars, however, predominate at locations far from sites (fig. 7-5). Between 700 and 1,200 meters from sites they account for 60.0

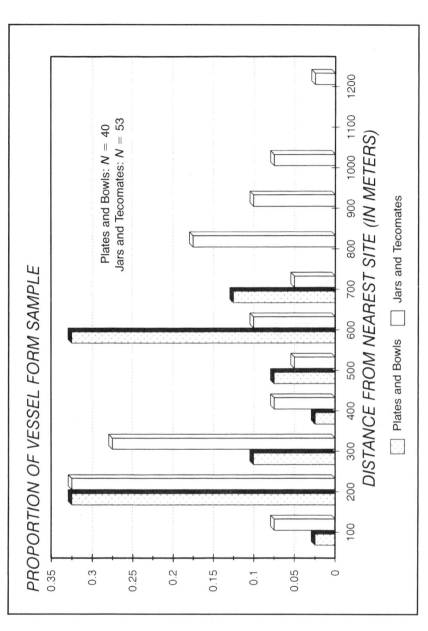

Figure 7-5. Histogram of Vessel Forms from Off-Site Squares

percent of the assemblage, but within 700 meters of sites *tecomates* are much less common (34.2 percent). Habitation sites and off-site contexts near them thus contain very similar assemblages of vessels, while those from contexts far from sites look quite different.

This difference in assemblage composition may relate to variations in labor organization. An off-site assemblage containing more bowls and plates might be expected in situations where land use involves teams of cooperating males, with one farmer (or his wife) bringing *tecomates* and jars filled with food and water to the field daily during the cultivation season and the others bringing their utensils (i.e., bowls and plates). This pattern, I have pointed out, characterizes off-site assemblages comparatively close to Formative villages. All other things being equal, the assemblage that would accumulate should be skewed toward service vessels, not jars and *tecomates*. The same assemblage should also occur at habitation sites where food was prepared for consumption by the entire residential group. Assemblages in which bowls predominate may therefore reflect group-oriented activities rather than tasks performed by individual males.

Alternatively, an assemblage rich in food preparation and storage vessels might be expected in contexts where agricultural plots were cultivated individually. Here, each farmer would have a bowl that might be cached at the plot overnight along with other field gear. Foodstuffs in a jar or *tecomate*, however, would be brought to the field each day by the cultivator. Inasmuch as the probability of a ceramic vessel breaking is related in part to the frequency and duration of use, more jars and *tecomates* should be discarded at locations far from communities, and because use was by individual farmers, not teams of cultivators, the assemblage should not look like those from habitation sites where pottery was used by members of the entire domestic group.

Data on vessel size support these inferences. Rim diameters for jars and *tecomates* are greater at off-site locations near villages than in contexts farther from sites, reflecting the use of vessels containing food and water for a greater number of persons. On the other hand, vessel sizes at distant off-site locations are smaller, implying individual use.

SPATIAL PATTERNING WITHIN AN EARLY FORMATIVE HOUSE LOT

Ethnoarchaeology has produced a growing body of information on household activity patterns and the material residues they generate. This research has involved inventories of household material technology (Deal 1983; Hayden and Cannon 1984) as well as studies of the organization of house-lot space and associated patterns of refuse disposal (Cook 1982; Hayden and Cannon 1983; Arnold 1986; Graham 1986; Killion 1987a). Two similar models of house-lot structure and refuse disposal have recently come to light as the result of this research.

The first model involves the notion of the Household Garden Residence Association (or house-lot model), as originally defined by Killion (1985, 1987a). The house-lot model is an ideal characterization of house-lot site structure, stipulating that how space is utilized in domestic contexts is related not only to the daily management of the lot in and around its major buildings but also to the frequency and variety of activities, some agricultural and others nonagricultural, that take place outside the lot. The house-lot model contains four components of archaeological relevance. The *structural core* consists of the lot's principal buildings such as the house, kitchen, and occasionally storage facilities that normally are the focus of archaeological survey and excavation. Around the core is the *clear area:* a zone that is meticulously kept free of debris because accumulations of household refuse in it interfere with routine domestic chores conducted around the core and outfitting activities for tasks performed outside the lot. In the Tuxtlas the clear area varies greatly in size. Households occupying lots with small clear areas derive most of their subsistence from outfields, while large clear areas more typically associate with households emphasizing intensive infield agriculture. Refuse that accumulates in this zone is swept or dumped in the *intermediate area*. Refuse is also dumped in the *garden-orchard,* which often accounts for more than 80 percent of the total amount of space within the lot.

The second model is based on research conducted by Brian Hayden and associates in the Maya highlands (Deal 1983; Hayden and Cannon 1982, 1983, 1984; Nelson 1987). This model is very similar to the house-lot model, though several important specific differences occur in gar-

167

bage disposal. The unit of analysis is the house compound, or *solar*, which is equivalent to the house lot in Killion's study. At the center of the compound are one or more residential structures, an external kitchen, and a few storage facilities that make up the *structural core* of the lot. These buildings are distributed around an *open-air patio*, which is constantly kept clear of debris by sweeping, material recycling, tossing, and child's play, all of which move refuse outward from the core. Much of this material ends up first in a *provisional discard area* (e.g., near walls, fences, and hedges) because of its "hindrance potential" with other household activities. Final discard occurs in the *toft zone* situated beyond the provisional discard area. Here garbage may be deposited in pits that have fallen into disuse, placed in refuse piles, or simply strewn over a large area.

Large amounts of refuse are also thrown into the street in front of the compound, carted to special dumps or nearby abandoned structures, or thrown into ravines and stream beds. Street dumping is more common in lower-density settlements where more space exists between house lots, whereas alternative disposal modes outside the compound are preferred in higher-density communities because street use is more intensive, neighborhood peer pressure is thus greater to keep arteries of transport clear, and ravines and stream beds are located nearby. Many compounds also have a *garden*, which judging from the size of the compounds may be quite large. In contrast to the house-lot model, little trash is dumped in a garden, though farmers recognize the value of organic refuse in maintaining soil fertility. Hard refuse is not placed in the garden because it damages agricultural tools. Different kinds of trash are also disposed in different ways. Glass sherds, for example, are generally placed in pits in the toft zone immediately or carted out of the house lot, whereas other less hazardous materials are left in provisional discard zones for longer periods of time.

The distribution of house-lot space in the house-lot model can therefore be conceptualized as a series of concentric rings extending outward from the residence. Each ring farther out from the core contains more trash until a threshold is reached in the garden or garden-orchard, after which the density of material falls off. Debris farther away from the core also tends to be larger—the result, it would appear, of missing smaller specimens while sweeping. The size of the house lot and various components within it appears to be related not only to the area of the garden or garden-orchard but also to the kinds of and intensity of activities conducted off-lot, especially the amount of wage labor, craft specialization, and the degree of reliance on infields and outfields as bases for household subsistence. The disposal of refuse also shifts with lot size, with trash concentrated at specific locations when less space is available and a broadcast approach preferred when lots are larger. The distribution of lot space in highland Guatemala can also be viewed as a series of rings of activity where refuse differentially accumulates. Trash generally increases in density with greater distance from the lot core, but very little hard refuse is placed in garden areas. Significant differences also occur in compound size that appear to be primarily a function of gardening as well as specialized activities that take place on the lot.

Information on Olmec house-lot structure at Matacapan comes mainly from a series of excavations at site TF-6. This group of excavations was located to the south of Mound 22 in an area of the site termed the "Teotihuacan Barrio."[3] The Teotihuacan Barrio consists of four large Classic period mounds plus several smaller ones that have been leveled by modern plowing, all located directly west of the main plaza (fig. 7-6). All four structures were situated around a small plaza constructed atop a low rise in the center of the barrio. The rise also supported one or more households of farmers during Early Formative times. Although no Olmec structures were discovered in the plaza per se (due to Classic period earth moving and the limited number of test excavations we consequently completed), the high density of Olmec materials and the distribution of trash and other features behind the mounds indicate that they were originally present. Such locations are often preferred for settlement because drainage after rainstorms is faster, inundation is less severe a problem, and trash disposal is facilitated downslope.

The area to the south of the rise contained a substantial amount of Olmec period refuse. A wide variety of different types of material was present, including large numbers of *tecomates*, the complete array of pottery service wares, anthropomorphic figurines, several jade artifacts, obsidian blades, cores, bifaces, and reduction debitage, some ground stone, and a few fragments of animal bone. The number and diversity of materials present suggest a domestic assemblage: in other

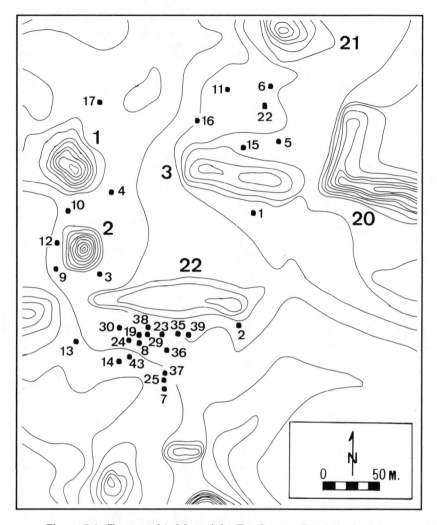

Figure 7-6. Topographic Map of the Teotihuacan Barrio in the Main Plaza of Matacapan

words, trash that was discarded from a household or group of residences situated nearby. A few bell-shaped storage pits were also present. These were located either close to the crest of Mound 22, where I suspect the residential core of the house lot was situated, or far from the mound, perhaps near an adjacent Formative structure. These tronco-conical pits are morphologically identical to Formative bell-shaped pits reported in other parts of Mesoamerica where they generally occur within ten to fifteen meters of houses (Flannery 1976; Santley 1977). All tronco-conical pits in Operation I-C were devoid of cultural materials, but they were completely filled with volcanic ash, suggesting use at the time of the volcanic eruption. Operation I-C did not uncover any Formative period burials; however, no excavations were placed near the top of the mound (due to permit restrictions). Thus, if the pattern of Early Formative interment was similar to that in Classic times (i.e., under and around the residence), no burials would be expected in the areas we exposed.

One of the most unusual features exposed in Operation I-C was the remains of an Early Formative agricultural surface. This agricultural feature was exposed in Pit 8 and had an undulating surface of ridges and furrows that ran to the southwest parallel to the slope of the terrain. Three ridges were present in the north profile of Pit 8 and parts of four in the south profile. The ridges were 72 to 110 centimeters apart, with the crests 12 to 22 centimeters above the swales, measurements that closely resemble the agricultural surfaces recently excavated in the Zapotitán Valley of El Salvador (Sheets 1982, 1983; Zier, this volume). No plant casts or carbonized remains were found on the ridges, suggesting that the vegetation had been cleared prior to the eruption. The deposit under the agricultural surface was 1.5 meters deep and contained large amounts of Early Formative pottery, implying use as a household garden or garden-orchard. Deposits near this surface also produced large amounts of trash, indicating that the area surrounding the feature had been used as a refuse dump. The agricultural surface had been covered by a deep layer of black volcanic ash. This same ash also occurs at other Formative sites in the Tuxtlas region such as Tres Zapotes (Chase 1981). Other Early Formative, ridged agricultural surfaces occur at Matacapan, but they are located hundreds of meters from the nearest habitation site. These features, I suspect, represent the remains of infields—an inference supported by the fact that they contain very little cultural material.

The distribution of trash on the house lot in Operation I-C shows some intriguing patterns. The density of Olmec ceramic materials on the surface of the deposit covaries with subsurface density ($r_s = 0.7386$; $Z = 2.9544$; $p < .004$), demonstrating congruence in the distribution of material in different parts of the excavation profile. The frequency of bowls is highly correlated with the frequency of jars and *tecomates* ($r = 0.9063$; $t = 8.3052$; $p < .001$), indicating similar modes of refuse disposal for these two classes of vessels. In contrast, total obsidian is uncorrelated with total ceramic refuse ($r = 0.0932$; $t = 0.3625$; $p > .2$), and obsidian density is not correlated with sherd density per level of occupation ($r = -0.0340$; $t = -0.1318$; $p < .2$). On the other hand, deposits with more total pottery tend to be deeper ($r_s = 0.7169$; $Z = 2.8677$; $p < .005$). This is particularly the case for squares with high densities of ceramics that average 85 centimeters in depth, whereas the deposit in squares with high densities of obsidian tends to be much more shallow (42 centimeters). Obsidian also tends to be found farther downslope from the top of Mound 22 than ceramics ($t = 6.803$, $p < .001$), implying disposal at greater distances from the structure. However, because the density of ceramics or the density of obsidian does not correlate well with distance from the top of the mound ($r = -0.1753$ and 0.2736 respectively), refuse disposal during the Early Formative period is not a simple negative linear function of proximity to the residence. Logarithmic transformation ($r = -0.1973$ and -0.0743) and rank-order correlation ($r_s = -0.3480$ and -0.3615) yielded similarly weak results.

Because the excavation with the agricultural surface produced the most Olmec pottery and had one of the highest sherd densities, distances from all other excavations relative to Pit 8 were recomputed. Recalculation of the correlation coefficient for density of ceramics produced an increase in the magnitude of the relationship ($r = -0.5620$; $t = -2.6315$; $p < .02$), implying a pattern of linear falloff in the density of material with increasing distance from the agricultural plot. Rank-order correlation yielded an even better coefficient ($r_s = -0.6930$; $Z = 2.7721$; $p < .006$). Transects across the excavated area show that the pattern holds best for those excavations northeast and southwest of Pit

8. The general pattern, then, appears to be one of increasing density of pottery on an alignment from the top of Mound 22 to Pit 8, followed by decreases in the density of material discarded. Early Formative deposits near Pit 8 also tend to be somewhat deeper ($r_s = -0.4253$; $Z = -1.7001$; $p < .1$), though the relationship is more complex than simple rank order. Obsidian densities, on the other hand, are not well correlated with distance from the agricultural surface ($r_s = 0.2298$; $Z = 0.9191$; $p > .5$), again indicating a difference in trash disposal patterns between ceramics and lithics.

In order to view the spatial patterning in the distribution of refuse, the ceramic densities were assigned to quartiles, and all squares with no ceramics were placed in the first quartile. As is readily apparent, squares close to and far from the top of Mound 22 contain little or no pottery, while squares in the immediate vicinity of Pit 8 all fall into the fourth quartile (fig. 7-7). Excavations spatially located between these extremes have low or moderate densities of ceramics. The high-density zone around Pit 8 suggests that some refuse was deposited in concentrated piles, whereas the zone of moderate density surrounding the area of high-density accumulation appears to be trending to the southeast: that is, in an arc around the mound. Squares with high densities of obsidian tend to occur in either the first or the fourth quartile ceramic squares, and squares with low densities of obsidian tend to occur in either the second or third quartiles. The distribution of obsidian is best described as punctuated: that is, consisting of small localized scatters. Sometimes levels with the greatest density of obsidian are situated below the top of the Early Formative occupation surface, suggesting intentional burial to avoid accidental injury from obsidian cutting tools after lithics had no functional use.

All the evidence presented thus far is consistent with the house-lot model. The area near Mound 22, especially its eastern side, appears to have been systematically kept clear of debris. Occupation levels under the Early Formative ash do occur, but they contain few artifacts.[4] The presence of a clear area adjacent to the top of the mound implies a residence situated there. It also suggests that the area around the structure was periodically swept clean of debris and that outfitting activities with respect to tasks performed off-site were regularly conducted there, in accordance with the house-lot model and supported by data

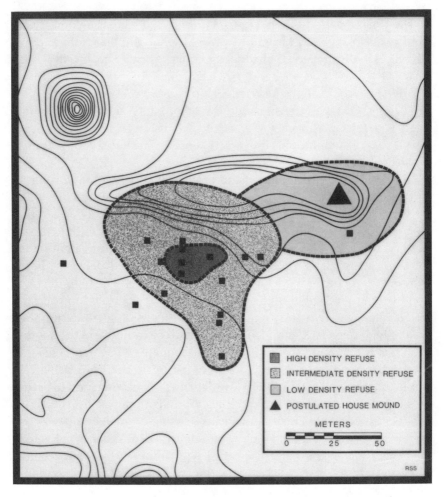

Figure 7-7. Early Formative Refuse Densities in the Teotihuacan Barrio

on the off-site distribution of Olmec ceramics. This clear area is adjacent to a transitional zone, where ceramic densities are low to moderate, and which is located farther downslope from the mound. Early Formative ceramics here are comparatively small in size and often are embedded in the occupation surface, probably from weathering, trampling, and/or pedoturbation after provisional discard. Farther away from the mound the Early Formative deposits are deeper and contain greater amounts of larger sherds. This locus of moderate-density accumulation indicates the presence of an intermediate or toft zone flanking the mound to the south where debris from the house-lot core was periodically swept or dumped.

The highest densities of ceramics occur in the garden or at the interface between the area around the ridged surface and the intermediate zone. The segment of the garden that was ridged was probably planted in maize, perhaps intercropped with beans and squash, while other economically useful species were probably cultivated in the parts nearby that were unridged. Household refuse was probably deposited in the garden to prevent its interference with other activities performed next to the house as well as to maintain soil fertility, which declines rapidly in the Tuxtlas without artificial enrichment after several years of repeated use. The garden may have also served as a zone of night-soil accumulation. Thereafter, refuse densities fall off dramatically; however, an increase occurs in the vicinity of Pit 13, located near the base of another small rise, indicating that another house lot was probably situated nearby.

Similar patterning was also evident at Loma Torremote, a Late Formative site in the Basin of Mexico (Santley 1977). Here, except for trash dumped in defunct tronco-conical pits located within the house, the structure contained comparatively little material. Small concentrations of obsidian artifacts, however, were embedded in house floors and patio surfaces, implying the presence of a few primary activity areas. Outside the house was an area relatively clear of debris, surrounded by a zone where refuse densities were much higher. Moreover, sherd size tended to increase with greater distance from the residence, suggesting that the clear area was also maintained by sweeping and that refuse disposal involved a provisional component. Although obsidian was also disposed of differently than ceramics, as at Matacapan, most trash ended up in subterranean bell-shaped pits located outside the house. At Loma Torremote very little material was found in the area I suspect was utilized as the garden. This lack of material may reflect hindrance to garden activities or more probably the ubiquity of tronco-conical pits, which could be filled with trash or which had to be topped with debris after falling into disuse.[5]

The degree of fit with the toft-zone model is also reasonably high.

The major area of disagreement is the high density of cultural material in and around the house-lot garden. Hayden and Cannon (1983) state that farmers prefer not to deposit large pieces of inorganic refuse on garden plots because it ruins the points and bits of gardening tools. As they point out, "one common strategy was to leave such debris in the toft area, close to the house until it was convenient to gather a load of it together and remove it as clutter refuse to a recognized dump within or outside the compound" (Hayden and Cannon 1983:146). The type of material to which they refer, however, includes stone pebbles and other similar kinds of debris. Such material, I point out, *is* very rare in the upper levels of the Formative garden area in Operation I-C as well as other agricultural features at Matacapan, but it is common in other Olmec contexts where no evidence exists of horticultural activity. Almost all of the material from food production contexts consists of potsherds and small pieces of obsidian. As a result, this refuse may not have been durable enough to have constituted a severe hindrance or hazard to gardening practices, or the need to maintain fertility may have outweighed the destructive effects that such refuse has on agricultural technology. It is also possible that differences exist in contemporary garden structure between the Tuxtlas and the Maya highlands, with hard refuse mainly dumped in garden-orchards that have a multi-tiered architecture in which fewer annuals are grown as stand crops. Because at least part of the Formative garden was ridged, implying the cultivation of annuals, this observation is difficult to support with the evidence from Matacapan.

If the Early Formative deposits exposed in Operation I-C are representative of other Olmec residential contexts at Matacapan, then zones with high densities of material do not occur next to structures. While plots of the density distribution of material will indicate the presence of a structure situated nearby, peaks in the distribution are more likely to show the specific location of trash dumps and small garden plots. Moreover, maintenance of the lot surface and the systematic removal of trash are likely to leave little evidence of primary activity contexts. Material residues in primary activity locations would only be expected in situations where the household regime produces microassemblages that become embedded in soft soil matrices. Most other materials, however, were probably subject to a variety of disposal and noncultural processes that greatly modified the distribution and composition of archaeological assemblages.

SUMMARY AND CONCLUDING REMARKS

In this chapter I have argued that inferences about Olmec land use can be defined in some detail if one adopts a spatial perspective that incorporates information from off-site locations as well as evidence of refuse disposal in ordinary household contexts. These data, I believe, are relevant to the question of prehistoric land use because patterning in their distribution reflects in part how households utilize space, both on-site and off-site. The archaeological record represents the accumulation of debris over long periods of time. Thus, rather than using the record as a vehicle to discuss the episodic behavior of human groups at particular moments in time, I have keyed into those properties of the record that reflect behavioral patterning over the longer term. Accurate description of the past is what we want; however, the perspective one may have to take and the level of description adopted may be radically different than those commonly provided by ethnographers. Data on short-term episodic behavior (e.g., the ridging of fields) should not be ignored when available, but archaeologists should not be hopeful that such evidence will be found except in the most extraordinary of situations: hence, my emphasis on off-site artifact distributions and household garbage disposal patterns as a means for describing elements of Formative economic organization.

The present study suggests the following observations about the organization of Olmec economy at Matacapan. Early Formative settlement patterns were dispersed, with evidence of only two levels in the settlement hierarchy; villages and hamlets. Villages may have been somewhat more differentiated than hamlets, but in the main they appear to have been very equivalent socioeconomically. That the landscape around Early Formative sites was utilized is supported by the fact that significant quantities of Olmec ceramics occur off-site. This assemblage of material patterns in systematic ways. More material in general is found closer to sites than far from them. Patterning also occurs by catchment area relative to habitation sites as well as by spatial zone at Matacapan, a fact that suggests major differences in the intensity of use and the organization of labor. The overall system of land use off-site appears to have been one involving the use of locations near sites as well as others farther away. The distance of these areas from habitation sites compares very favorably with the distance walked today to off-site plots, indicating an infield land-use system.

Excavations in areas of midden accumulation imply that gardens or garden-orchards were another component to Formative subsistence. Gardening is indicated by the presence of a ridged cultivation surface in association with deposits of Early Formative trash, suggesting the disposal of household refuse to enrich the plot, in congruence with the house-lot model. Garbage disposal during Early Formative times appears to have been very systematic, involving the maintenance of clear areas, both casual and provisional discard, and differences in how different classes of material were dumped, as reflected in the distribution of ceramics and lithics in the assemblages from different contexts. Other ridged surfaces occur at Matacapan, but these features probably represent infields, not gardens or garden-orchards, to judge from their distance from habitation sites and the paucity of ceramic materials found in them. A substantial part of Early Formative Matacapan therefore consists of a low-density scatter of material that reflects the utilization of off-site localities, as I have argued for agricultural purposes. The size of the garden area exposed is difficult to determine with certainty. In all likelihood the plot was planted in a variety of cultigens, including annuals such as maize, beans, and squash, as well as perennials grown nearby.

Although many of the implications of the house-lot and toft-zone models show up in the archaeological record at Matacapan, the degree of fit is not perfect. For example, while the disposal of obsidian is unrelated to that of ceramics, obsidian still constantly occurs in all samples. In contrast, the pattern today involves the disposal of glassy material in pits or dumping in out-of-the-way places. Thus, in the Early Formative period, obsidian appears to have been discarded in a somewhat more carefree manner than glass is disposed of in contemporary house lots.

The reason for this difference in disposal patterns may be tied to the size of the specimens being discarded and the amount of edge damage manifest on them: in other words, their hindrance or hazard potential. Almost all of the Olmec obsidian assemblage consists of extremely small pieces of reduction debitage, snapped prismatic blades, and percussion microblades, whereas modern assemblages are dominated by much larger fragments of glass from beer bottles and other containers. Although a glass sherd will produce a cut no matter what its size, the depth and severity of the puncture (i.e., the hindrance potential) is related to the size and thickness of the specimen and the sherd's orientation in the ground prior to incision. Most of the archaeological assemblage also bears evidence of heavy utilization. At Matacapan many Formative obsidian tools appear to have been recycled for other uses after they were no longer serviceable as cutting implements. Dulled use edges are consequently very common, even on small specimens. On the other hand, modern glass sherds are frequently discarded immediately after breaking to minimize the danger of cut and puncture wounds because fresh fractures are still extremely sharp. The hindrance potential of glass objects, then, should vary systematically in terms of the uses (or reuses) to which they are put as well as their relative sizes, which might affect variations in disposal patterns. Disposal in both contexts might still emphasize the immediate removal of glassy materials from cultural living surfaces, but households dumping small pieces of dull material might be somewhat more casual in their discard behavior.

The strategy that I have adopted in this chapter is one involving an argument of enumeration. Thus, patterns from the contemporary world have been "fitted" to evidence from the past in an effort to show that the specific archaeological case can be viewed as a special instance of a more general pattern definable in the present. No independent evidence has been offered nor have other consequences of the model been derived. As a result, no tests of alternative models of land use and refuse disposal have been made to increase our confidence that the model applied is the most appropriate characterization of Olmec economy at Matacapan. If alternative models are to be rigorously evaluated, then more research must occur in the following areas.

First, we need more information on other factors affecting variability in refuse disposal patterns in household contexts. Neither of the house-lot studies deals with the effects of craft specialization. Do potters producing ceramics in different types of contexts (e.g., household industries versus workshop industries versus manufactories) discard refuse in a similar manner? In other words, what effects, if any, do specialization and the scale of specialized production have on the organization of space and garbage disposal patterns? Moreover, does systematic variability occur by class of industry? For example, industries employing additive technologies (i.e., ceramic production) might be expected to be organized very differently from industries utilizing subtractive technologies (i.e., obsidian working) because the latter produce vast amounts of hazardous waste. Finally, does variability occur by socioeconomic level? All of the studies discussed in this chapter deal with peasant households of rather low economic status. Elites, in contrast, might be expected to behave very differently. Households in highland Guatemala rarely dig pits to dispose of their refuse. Rather, they take advantage of alternative disposal modes or fill pits with garbage that were originally dug for other purposes. Persons of high status, however, might have the economic clout for disposal in pits because such tasks would be handled by servants or clients of low rank, not by the elites themselves. I am not arguing that such was the case, only that we do not know that it was not.

Second, there is a paucity of information on variation in agricultural intensity, the kinds of work gangs operative off-site, and their impact on community settlement patterns. As Killion (1987a) has pointed out, settlement patterns vary greatly in the Tuxtlas today. This variability in part is a function of environmental differences, combined with variations in agricultural intensity, the location of intensively cultivated plots relative to villages, the degree of reliance on gardens or garden-orchards, and proximity to regional centers or lines of transit leading to them. Variation of this sort may have dramatic impact on the organization of agricultural labor (e.g., cooperative versus individual), which in turn could condition the use of space on-lot as well as the kinds of materials discarded off-site. Besides agricultural implements per se, we also have very little data on the other kinds of technology taken off-site. All studies of contemporary land use in Mesoamerica describe field preparation, planting, cultivation, and harvesting technology. Such implements, however, are very rare in off-site contexts in the Tuxtlas if the data from Matacapan are representative. More common are fragments from ceramic vessels, both bowls and jars. At Matacapan this technology occurs in association with agricultural features and reflects, I believe, provisioning for agricultural tasks conducted off-site. How this provisioning varies in terms of the variability discussed above is regrettably unknown. Likewise, we lack information on off-site disposal patterns. If disposal patterns in systematic ways on-site, then it seems reasonable to assume that it may also pattern in predictable ways off-site. What I am suggesting is that the same kinds of ethnoarchaeological studies undertaken in house-lot contexts ought to be conducted using the agricultural field as the unit of analysis.

My original purpose in writing this chapter was to show that knowledge about Olmec economy can be accumulated through the systematic analysis of archaeological materials from ordinary contexts. I have also pointed out that present knowledge about the factors that condition the distribution of archaeological assemblages is insufficient to test models of prehistoric economic organization. Although the results presented above are highly suggestive, they are hardly demonstrative, at least not at Matacapan for reasons indicated above. Our continuing work in the Tuxtlas, however, suggests that research that combines information on modern land use, refuse disposal, and house-lot structure, their material correlates, and archaeological patterning is a profitable line of inquiry. My knowledge of the problems with which archaeologists are grappling in the humid tropics indicates that it may be useful in other areas of Mesoamerica as well.

ACKNOWLEDGMENTS

I wish to thank the people of the modern communities of Matacapan, Sihuapan, Caleria, Comoapan, San Andres, and San Isidro Texcaltitan, especially Lic. Miguel Turrent, Lic. Carlos Silva, Lic. Miguel Castillo, and Carlos Turrent, for their assistance throughout various stages of our research in the Tuxtlas. Special thanks are reserved for Robert H. Cobean and Ponciano Ortiz Ceballos, project ceramicist, for keeping the Matacapan project afloat in its formative years, as well as all of the archaeologists who worked as crew chiefs and volunteers over the years. I also thank Angel Garcia Cook, Joaquin Garcia Barcena,

Daniel Molina Feal, and Daniel Nahmad for expediting permit requests and facilitating our research. Research in the Tuxtlas region was conducted with the permission of the Instituto Nacional de Antropología e História in Mexico City. During the course of the Matacapan project Robert S. Santley acted as general project director, with Richard A. Diehl as field director in 1982 and 1983 and Christopher A. Pool as field director in 1986. Information presented in this chapter is based primarily on data collected during the 1982, 1983, and 1984 field seasons. Evidence from the 1986 field season has been included when appropriate. This research would not have been possible without grants from the University of New Mexico, the Tinker Foundation, the Mellon Foundation, and particularly the National Science Foundation (BNS-8120430, BNS-8302984, BBS-8403810, BBS-8412175, BNS-8505041, and BNS-8520615).

NOTES

1. The surface survey strategy employed by the Matacapan project was multistaged. The area to be mapped was extensively surveyed first to define the universe to be sampled and to obtain general information on site layout and other sources of variability. An intensive survey was then conducted. The intensive survey employed a systematic collection design. Each area of the site was divided into an array of transects 3 meters wide, transects to be sampled spaced about 75 to 100 meters apart were then selected, and a 3-by-3-meter surface collection was obtained every 10 meters along each transect. After vegetation removal, controlled surface collections were retrieved from all squares in the sampling frame regardless of the density of materials. Transects laid out in this fashion often missed concentrations of surface material or mounded architecture. In these cases secondary transects were run off the main axis to obtain samples from buildings and refuse dumps. Four specialized ceramic production areas were later selected for additional survey. Three of these areas also had Olmec occupations at them. This resurvey utilized probabilistic procedures as a basis for unit selection. A grid of units 9 meters square was arrayed over each area, units to be sampled were drawn randomly, and surface collections were taken from each 3-by-3-meter square within each unit. Altogether the survey produced more than 6,500 collections from an area 12.67 square kilometers in size. Although some parts of the site were sampled more

intensively (e.g., Comoapan, Mound 61, and the Teotihuacan Barrio), the overall sampling intensity was about 0.46 percent.

2. This estimate was derived as follows. First, the area of each site was computed and that figure was divided by the average house-lot size for farmers in the Tuxtlas today (0.2196 hectares, derived from Killion 1987a) to obtain an estimate of the mean number of households expected. This total was then multiplied by the average population of those households (6.8 persons) to get an estimate of the total number of persons represented. In order to estimate a likely range, household population and house-lot size were set equal to their respective means plus or minus one standard deviation and the site totals were summed. This method produced a mean of 5,561 +/−4,652 persons, or an average population density of 438.91 persons per square kilometer. However, because house lots in dispersed communities in the humid tropics are often abandoned after several generations of consecutive use, lot lifespan was assumed to be fifty years, which reduced the Early Formative mean to 927 +/−775 persons, or an average population density of 73.2 per square kilometer. Even if probable momentary population equaled the mean minus one standard deviation, rural population density still exceeds 12 persons per square kilometer. While hardly an enormous population relative to later time periods, this figure is fairly high for rural settings during the Formative period in Mesoamerica, implying that agriculture involved an intensive component.

3. My use of the term *Teotihuacan Barrio* is a misnomer. While originally it appeared that Teotihuacan-style materials were confined mainly to this locus, subsequent work has demonstrated that they occur throughout Matacapan. The use of this term is consequently for geographic purposes only.

4. A few extremely small sherds were found in all Olmec levels, but because they could not be reliably classified, these materials were deleted from the tallies.

5. Altogether, more than seventy tronco-conical pits were excavated at Loma Torremote. Although many contained considerable amounts of refuse, including large sherds and pieces of manos and metates, others yielded very little material. This variation suggests that these pits were not originally dug to dispose of household refuse. Rather, they probably had a storage function and were quickly filled with debris from about the compound after the pits were no longer used for that purpose. If a lot of trash occurred in provisional discard locations, it was probably thrown into the pit, but if little household refuse was available, then the pit was filled with dirt, pieces of daub, or whatever material was on hand.

CHAPTER 8   PATRICIA A. MCANANY

Agricultural Tasks and Tools

PATTERNS OF STONE TOOL DISCARD NEAR
PREHISTORIC MAYA RESIDENCES BORDERING
PULLTROUSER SWAMP, BELIZE

THE PAST TWO DECADES have brought about a revolution in our ideas regarding the agricultural support base of ancient Maya society. It is an overstatement, however, to say that we have arrived at an understanding of the myriad techniques employed by the Maya to fine-tune and enhance regional agricultural production. In fact, the "revolution" to date has pivoted on the identification of relict agricultural features (terraces, raised fields, ridged fields, and field walls) through aerial reconnaissance and photography and through more confirmatory field survey/mapping and excavation (Adams, Brown, and Culbert 1981; Adams et al. 1990; Bloom, Pohl, and Stein 1985; Culbert, Levi, and Cruz 1989; Freidel and Sabloff 1984; Freidel and Scarborough 1982; Gliessman et al. 1985; Hammond et al. 1987; Harrison 1977, 1978; Harrison and Turner 1978; Healy et al. 1983; Lambert and Arnason 1983; Pope and Dahlin 1989; Puleston 1976, 1977, 1978; Scarborough 1983; Siemens 1982; Siemens and Puleston 1972; Turner 1983a; Turner and Harrison 1983, among others). Some of the best confirmatory evidence of agricultural intensification in the form of swamp reclamation and field raising comes from Pulltrouser Swamp; this chapter is a complementary study of locational patterns of stone tool and debitage discard near prehistoric Maya residences bordering Pulltrouser Swamp.

Attention to the diversification as well as intensification of Maya

AGRICULTURAL TASKS AND TOOLS

agriculture (Harrison and Turner 1978) has effectively overturned earlier notions about the monolithic swidden base of Maya agriculture (Turner 1978); therefore, we can now turn to a consideration of the changes in the organization of agricultural labor that must have been attendant upon this agrarian transformation. In this study the nature of the changing agricultural tasks is suggested by reference to ethnographic case studies. These examples pointedly indicate that, contrary to popular opinion, it is not the construction of agricultural field features that garners an increasingly large proportion of agricultural labor but rather tasks, such as weeding and tilling, that are associated with field maintenance. These tasks require no technological revolution in agricultural implements but do, in fact, affect the manner and frequency with which tools are used.

Having established a linkage between the organization of agricultural labor and patterns of agrarian tool wear as well as locations of breakage and refurbishing, I then move to a consideration of the Pulltrouser Swamp agricultural system and propose a model of land use that involves an initial system of infield-outfield plots. During the Late Formative or Early Classic periods, however, I propose that a constriction and intensification of agricultural strategies resulted in increased emphasis on what Killion (this volume) has referred to as "settlement agriculture." Field raising along the borders of the swamp represent just one highly visible component of settlement agriculture; the extremely fertile lands between platform clusters were also extremely important cultivation spaces.

This model is examined in light of the locational patterns of tool fragments and of debris from refurbishing ancient agricultural tools used in the vicinity of Pulltrouser Swamp. Temporal trends in the lithic data provide some substantiation for the model. The lithic assemblage utilized in this analysis was recovered from residential and interresidential contexts at Pulltrouser Swamp, Belize (fig. 8-1).

INTENSIFICATION OF PREHISTORIC TROPICAL AGRICULTURE

The term *agricultural intensification* is used throughout this chapter in reference to shortening of the fallow period, as defined by Boserup

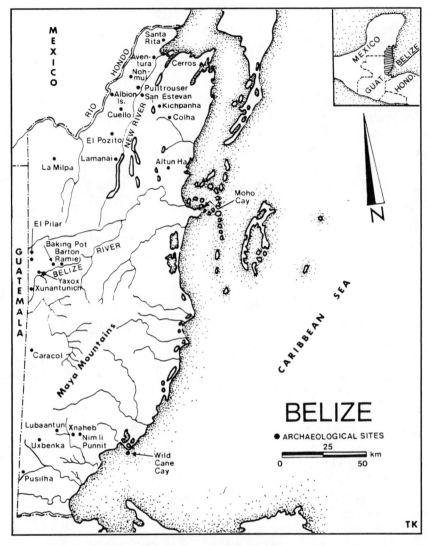

Figure 8-1. Belize, Central America, with Archaeological Sites and Other Place Names Discussed in Text Indicated

AGRICULTURAL TASKS AND TOOLS

(1965:43), and increased inputs of capital, labor, and skill on marginal lands (as used by Brookfield 1962, 1972; Brookfield and Hart 1971; and Geertz 1963; see also Turner and Sanders, this volume). The construction of agricultural field features such as terraces and raised fields is a form of increased capital and labor input; therefore, land subjected to this type of modification is generally fallowed less frequently than unmodified land. A combination of these two conceptual definitions of land-use intensity (as per Turner and Doolittle 1978) seems to encompass the important variables of intensification. The reclamation of wetlands, hillslopes, or rocky terrain—that is, the expansion of an agricultural system into secondary habitats (vertical expansion as per Hammond 1978:26)—is considered to be a form of intensification. Both ethnographic (Netting 1968) and archaeological (Wiseman 1983) evidence suggest that these "marginal" lands are reclaimed *after* production on optimal lands has been maximized (Farrington 1985a:5). In effect, the archaeological remnants of raised fields and terraces documented in the Maya lowlands are the endpoint of a long-term reduction in intensification options.

Space does not permit an extended discussion of the systemic causes of intensification. In the case under study, three variables (local population growth, increased taxation demands, and participation in local exchange networks) appear to have played a major role in this trajectory. Any or all three combined could have stimulated intensified agricultural practices. Population growth (as measured by the construction or renovation of residential platforms) proceeded more or less hand-in-hand with the expansion and intensification of the local agricultural system (Turner 1983b:21; Turner and Harrison 1983:248; McAnany 1986) with parts of the local community around the swamp experiencing rapid settlement expansion, particularly during the Early Classic (Walling and Minc n.d.). At some point during the Late Preclassic or Early Classic periods, these small, semiautonomous communities ringing Pulltrouser Swamp probably were incorporated into a state-level political entity and the forces of political economies, specifically taxation demands, undoubtedly began to play a persuasive role in agrarian change. Finally, through technological analysis of stone tools and debitage, the essentially "consumer" role of these communities in a local exchange network has been established (McAnany 1986, 1989;

Shafer 1983). The specifics of their "production" role has yet to be established; however, agricultural "exports" (as suggested by Freidel and Scarborough 1982:153) are a likely candidate and also may have played a role in stimulating field expansion.

The process of agricultural intensification, especially in temperate latitudes, has been linked historically to a change in agricultural tool technology (Childe 1937:138–39); specifically the progression from dibble stick to hoe and ultimately to the use of a plow and traction animals (see Boserup 1965; Smith and Young 1972; and Turner and Sanders, this volume). In the Precolumbian Americas, however, traction animals were completely unknown, and it is only in South America that a type of footplow *(chaqui-taclla)* has been recorded for the protohistorical period (Donkin 1970). Likewise, stone tools recognizable as hoes (with a broad blade amenable to a transverse hafting) have a patchy distribution during protohistorical times and earlier. Hoes existed in North America (for example the well-known Mill Creek and Dover chert hoes of the Mississippian period; Koldehoff 1987:163) and in South America where they are illustrated in early historic documents (Guaman Poma de Ayala 1936). On the other hand, this specific tool form is not well represented (in stone) in lowland Mesoamerica. Carneiro (1961:62) states that the hoe was absent from the entire circum-Caribbean area and the tropical forest. In a historical review of Precolumbian field implements in Mesoamerica and South America, Donkin (1970:507) concurs with Carneiro and states that the simple fire-hardened digging stick (sometimes tipped with stone or metal or weighted) and the *coa* (a stick with a flat, flared wooden base) seem to represent the full inventory of Precolumbian field implements in Mesoamerica. In a broad historical sense, the paucity of agricultural implements traditionally associated with agricultural intensification in Euroasian settings has lent tacit confirmation to the "swidden thesis" (Turner 1978) or the notion that contemporary Maya agriculture supplied an appropriate analog for the past.

In the New Guinea highlands, Golson (1977) has grappled with the commonly accepted equivalence between a typological progression of agricultural implements and agricultural intensification. The tropical agriculturalists of highland New Guinea are notable for supporting highly dense populations with intensified techniques of sweet potato cultivation (including swamp reclamation). Their inventory of agricultural implements, however, seldom included more than a wooden spatulate hoe and digging stick and a stone-bladed ax (Golson and Steensberg 1985). Thus, the following questions emerge: Does complex agriculture necessarily require complex tools? If tool technology is not diagnostic of agricultural intensity, then what is?

Insight to these questions can be gained by attention to the ethnography of the Kapauku of Irian Jaya. Pospisil describes their intensive drained-field agriculture as follows: "The main aspect of the complex cultivation of Kapauku is the preoccupation with the soil itself. This is not only drained of surplus water and fertilized by green as well as decomposed (composted) plant material, but it is also turned over with a spadelike tool in order to cover the layer of fertilizer and to form beds raised well above the surrounding area" (1963:122). Herein may lie one of the fundamental distinctions between prehistoric agricultural intensification in tropical as opposed to temperate latitudes. In tropical latitudes, intensification can involve the improvement and enrichment of the microenvironment of cultivation rather than the development of new agricultural implements. To a great extent this emphasis may be due to the greater susceptibility of tropical soils to chemical leaching and erosion (Fittkau and Klinge 1973; Nye and Greenland 1960; but cf. Turner and Sanders, this volume) rather than to any inherent infertility of soils in tropical latitudes.

Viewed from this perspective, we can anticipate that agricultural intensification in tropical areas may result in significant changes in the chemical and nutritive properties of the soil but not necessarily in the introduction of new tool forms. This situation would pertain, in particular, to areas such as the Maya lowlands that lacked domesticated animals that could be used for pulling plows and as a source of fertilizer. On the other hand, even though the agricultural implements may not change, the labor associated with the cropping of a cultivation plot does, in fact, change dramatically as a plot is repeatedly cropped. This change in agricultural tasks or the organization of labor is a constant of agarian change. In short, agricultural intensification may follow at least two pathways: (1) reorganization of labor using the same tools, or (2) reorganization of labor using different tools. Investigation of changes in the composition of agricultural labor, therefore, is directly relevant to understanding ancient Maya agricultural economy.

TASKS OF CONTEMPORARY TROPICAL AGRICULTURALISTS

Due to the archaeological visibility of agricultural features such as raised fields and terraces, considerable attention has been focused on the quantification of the labor involved in their construction (e.g., Turner 1983b). The construction event, however, is of limited temporal duration and, while perhaps requiring periodic rebuilding, is essentially nonrepetitive. The bulk of agricultural labor is actually expended in the routine maintenance of frequently cropped plots. This latter task involves highly repetitive work that usually includes some combination of weeding, mulching, and soil tilling.

Changes in the relative proportion of these agricultural tasks under systems of extensive as opposed to intensive cultivation can be examined by reference to ethnographic studies of time and labor allotments. Labor demands associated with weeding, in particular, have often been discussed in the context of plot abandonment among shifting agriculturalists. A few authors have attempted to quantify the magnitude of this problem during the latter years of a plot's productivity (Carneiro 1961:57; Clarke 1971:164; Johnson 1983:51; Kelly and Palerm 1952:100; Morley and Brainerd 1956:148). Within the context of shifting cultivation, labor input studies show that significantly more time is spent in weeding fields that are more than one year old or are planted in an area of cleared secondary growth. In a calculation of unit hours per hectare, Conklin (1957:150, table 15) shows that the Hanunóo invest twice as much time in weeding when a swidden plot is sown in secondary growth than in a new field (300 hours per hectare in climax forest and 600 hours per hectare in woody or bamboo second growth).

In another study of labor input Bergman (1980) analyzed the amount of labor (hours/hectare) expended by the Shipibo of the Amazon rainforest. The Shipibo are extensive agriculturalists, subsisting chiefly on bananas and on protein from hunting and fishing. They cultivate plots in several different locations: naturally fertilized mud bars, levees, and

AGRICULTURAL TASKS AND TOOLS

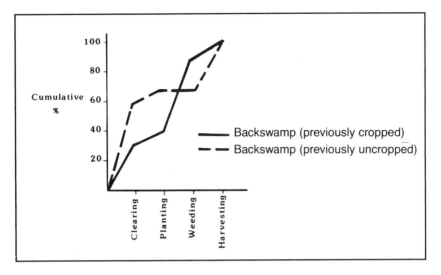

Figure 8-2. Shipibo Agricultural Tasks Differentiated by Field Type (raw data from Bergman 1980:29 converted from hours/hectare)

backswamps. The amount of labor expended during clearing, planting, weeding, and harvesting at two contrasting backswamp locations (previously cropped and uncropped) are shown in figure 8-2. These data are presented in a cumulative frequency form in order to direct the focus of this analysis not to changes in the total amount of energy expended but to the shift from field *preparatory* to field *maintenance* tasks that accompanies repetitive cultivation of a plot.

The Shipibo data indicate that clearing a previously uncropped backswamp plot consumes about 60 percent of the total labor input. At previously cropped locations, on the other hand, only 30 percent of the total labor input is consumed in clearing. The two practices are highly distinctive in their cumulative increase in labor time from planting to weeding. Little or no weeding is necessary in the previously uncropped backswamp; in the previously cropped backswamp, however, weeding consumes 46 percent of the total labor input!

The implications of this pattern—marked increase in field maintenance tasks with repetitive cropping—can be strengthened by reference to another ethnographic example in which a similar relationship

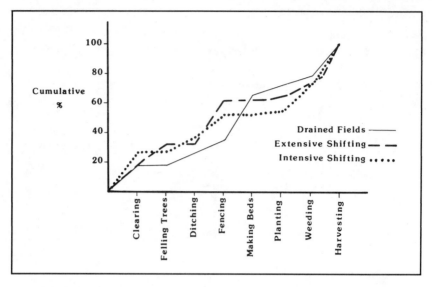

Figure 8-3. Kapauku Agricultural Tasks Differentiated by Field Type (raw data from Pospisil 1963:422 converted from hours/900 square meters)

exists. The Kapauku of Irian Jaya practice three forms of cultivation: extensive shifting, intensive shifting, and drained-field agriculture. Pospisil (1963:442) observed the number of hours expended per 900 square meters of garden area for eight components of cultivation: clearing underbrush, felling trees, digging ditches, building fences, making beds (for drained fields only), planting, weeding, and harvesting. These data are also presented as cumulative percentages (fig. 8-3) and are not dissimilar from those of Bergman.

The two forms of shifting cultivation (extensive and intensive) yield similar work profile patterns with one outstanding dissimilarity. Under intensive shifting cultivation, less or no time is spent felling trees but substantially more time is spent weeding. In contrast, agricultural tasks involved in the initial construction of drained fields create a work profile that is very distinct from shifting cultivation in several ways. First of all, tasks preparatory to creating the cultivation plots or beds consume less than 40 percent of the total labor output. Ditching, fencing, and bed preparation (fertilizing, hoeing, final ditching, and the careful arrangement of sweet potato vines) consume the lion's share of the labor input (39 percent of total task time). The task of weeding, which is a lengthy task in intensive shifting cultivation, decreases from 20 percent to only 5 percent during the initial preparation of drained fields. This decrease in weeding is due to the fact that the soil is completely turned over during hoeing and, at this time, the weeds are pulled up by the roots. Pospisil (1965:8) notes that, between plantings, these plots are fertilized with nutrient-rich "washed-down mud" from the bottom of the ditches and that, as a result, cultivation can "continue almost indefinitely." Fertilizing and weeding, therefore, probably consume a large proportion of the labor expended during the later croppings of these drained fields.

These examples demonstrate that as agricultural plots are repetitively cropped the proportion of total labor time spent in preparatory activities decreases and that spent in maintenance activities, such as weeding or hoeing, increases. With this relationship in mind, it seems worthwhile at this time to examine more closely ethnographic accounts of the mechanics of weeding and the tools of hoeing. We anticipate that agricultural tools used in such capacities will be subjected to a highly redundant use regime and that they will bear traces of such use in the form of specific patterns of wear and breakage that are relevant to understanding the function of stone tools recovered from domestic and field contexts around Pulltrouser Swamp.

Weeding: Implements, Mechanics, and Patterns of Wear

Weeding usually consists of some combination of systematically pulling weeds up by the roots or severing the stalk at, or slightly above, ground level. Although pulling the weeds up by their roots more effectively reduces the competition for nutrients and makes for a cleaner garden plot, some researchers maintain that the forest succession (in a swidden plot) is disrupted by pulling tree seedlings up by the roots (Kelly and Palerm 1952:114). Eventually, as the swidden plot matures and is left to fallow, grass may invade the plot because the natural forest succession has been protracted or aborted by eliminating the tree seedlings. This very pattern was observed on experimental *milpa* plots at Chichen Itza (Morley 1946:148). Given the absence of steel machetes in prehistoric times, Morley assumed (incorrectly) that no functional equivalent to the machete existed and, therefore, asserted that the ancient Maya must have weeded their plots by hand pulling the weeds up by the roots. Thus, he concluded that grass incursion may have been a limiting factor "to the indefinitely continued use of a plot" (1946:148).

Regardless of the imputed relationship between pulling up weeds by the roots and grass incursion, Johnson (1983:50) reports that the Machiguenga of the Amazon Basin weed almost exclusively by hand and rarely use their machetes. In contrast, Cowgill (1961:24) notes that the Maya of the Lake Peten region cut weeds with a machete, either at ground level or fifteen centimeters above ground level. Both the Peten Maya and the Machiguenga practice shifting cultivation. In Cowgill's study, the average cycle for the Peten Maya is four years of fallow after a single crop or six to eight years of fallow after two crops (1962:276). In contrast, the Machiguenga fallow for twenty years after three crops. The extended fallow period practiced by the Machiguenga may explain why they are not concerned with the problem of protracted forest succession.

The Totonac of the central Gulf Coast of Mexico weed their plots using a *coa* by holding it at a very sharp angle to the soil and cutting the weeds while at the same time stirring the soil slightly (Kelly and Palerm 1952:113). The modern *coa* described by Kelly and Palerm has a broad, flat metal blade hafted onto a long wooden handle in the same axis as the blade. Thus, although the *coa* is not hafted or shaped like a hoe, it is used to cultivate the soil around plants.

Among the drained-field cultivators of highland New Guinea, the Kapauku female, who is almost solely responsible for weeding, uses a narrow, wooden, paddle-shaped tool with which she breaks the ground, severs the roots of herbs and grasses, and then pulls them out (Pospisil 1963:98). This time-consuming process appears to have been a limiting factor to the amount of land under cultivation. That is, Pospisil (1963:98) found an extremely high positive correlation ($r = 0.8$) between the number of adult females per household and the extent of area under cultivation. This correlation suggests that, among intensive agriculturalists, the labor demands of field maintenance are so great that mustering a task force (male or female) that is large enough to meet these labor demands may be the single most important factor in maintaining or expanding field areas. Pospisil's well-documented relationship between agricultural holdings and household size is important in terms of understanding the Maya propensity toward large, agglutinated residential units.

Conklin (1957:105) gives an excellent account of the mechanics of weeding among the Hanunóo with detailed information regarding the angle at which a weeding tool is held and the nature and depth of contact between the tool and the soil. Women, who do two-thirds of the weeding, use a blunt-ended knife that is twenty to twenty-five centimeters long (Conklin 1957:104), and they weed by manual extraction, uprooting, jab cutting, or side cutting. To uproot a plant, the knife is jabbed into the ground at an angle in front of the stem base and then twisted so that the plant is pried loose. Jab cutting is used for large plants or small weed clumps; the plant is severed just above the ground level by the broad, chisellike tip of the knife, which is held parallel to the ground surface. Large weed clumps are removed (two to five centimeters from ground level) by a single sidestroke of the blade (Conklin 1957:105).

Further evidence of the blade coming into contact with the soil during weeding is provided by Bergman's (1980:98–99) description of the Shipibo of the Amazon Basin. Among the Shipibo, weeding the banana *chacras* is a male activity. Using a machete and a *gancho* (meter-long, hooked sapling), the male cuts the weeds at or slightly below ground level with a horizontal stroke of the machete and pulls the weeds out of the way with the *gancho*. Because the soil is fine-grained and stone-free, the men apparently do not worry about dulling their machetes (Bergman 1980:99).

When grasslands are being cleared for cultivation even greater contact occurs between the soil and the cutting tool because more effort is expended in breaking up the roots of the grass mat. As Shipibo women clear the red *gramalote* grass from mud bars in order to plant maize, they actually insert a short-bladed machete fifteen to twenty centimeters into the mud in order to cut the thick grass roots well below the surface (Bergman 1980:101). A nearly identical account of the mechanics of "grass pulling" by female Tiv work groups is given by Bohannan (1954:16).

This selective survey of ethnographic accounts of weeding suggests that some parameters of this labor task may be applicable to an examination of the archaeological record. Specifically, the tools used for weeding, the mechanics of weeding, and the frequent contact between the soil and the weeding tool have clear implications for tool wear patterns. These implications may be summarized as follows.

(1) All weeding tools are described as knives, machetes, or thin wooden spades; in other words, weeding tools are thin and lack the hefty mass of a metal or stone ax.

(2) In six of the eight accounts discussed above the weeding tool is in contact with the soil; in five accounts it is inserted into the soil; and in one account it is used just above and parallel to the ground surface. Through time and between edge refurbishings such a weeding tool can be expected to bear evidence of this contact in the form of nicking, if stones are present in the soil, or polish, if the soil is a mulched, fine-grained, or high-silicate clay, stone-free matrix such as existed in the raised fields of Pulltrouser Swamp.

(3) The weeding method labeled uprooting, which is analogous to the Totonac method of weeding/hoeing with a *coa*, involves a steep angle of contact between the weeding tool and the soil. In a stone-free matrix this constant contact between the tool face and the soil would result in facial polish and striations (angled oblique to the working edge of the tool) over a large portion of the surface of the blade. With this discussion of the tasks and tools of contemporary tropical agriculturalists in mind, we now turn to the examination of prehistoric agriculture at Pulltrouser Swamp.

AGRICULTURE AND STONE TOOL USE AT PULLTROUSER SWAMP

Located in the small Central American country of Belize (fig. 8-1), Pulltrouser Swamp is not a typical Maya center with a core of monumental pyramidal architecture encircled by residential structures but rather is a swampy depression covering 8.5 square kilometers around

AGRICULTURAL TASKS AND TOOLS

which dispersed and very modest residential dwellings were constructed. Spatial clustering of dwellings is due, in large part, to natural features; nevertheless these clusters have been given individual names, and their locations are identified on figure 8-4; Tibaat, Chi Ak'al, Pech Titon, K'axob, and Kokeal (Harrison 1988, 1990). The community of Yo Tumben, which is located to the south of Kokeal and along the New River, has only been partially investigated and appears to be primarily Preclassic (Harrison 1983:155). Only Kokeal and K'axob contain large plaza groupings with pyramidal architecture (Harrison 1983; McAnany n.d.).

Pulltrouser Swamp drains the east flank of the highest, westernmost ridge in the relatively flat, northern portion of Belize. Several Classic period Maya centers (with probable or demonstrable Preclassic nuclei) are located in proximity to Pulltrouser Swamp and, in fact, leave very little breathing room for the expansion of either settlement or agriculture. San Lorenzo and San Luis are on top of the ridge less than 5 kilometers to the west (fig. 8-4). The central precinct of Nohmul is only 2.5 kilometers from the northwestern edge of the swamp complex, and much of the intervening space is covered with residential structures (Pyburn 1987, 1989). Nohmul is the largest political center in northern Belize (Hammond 1985), and its growth during the Classic period assuredly affected the agrarian structure of its immediate hinterland. In a southeastern direction, Pulltrouser Swamp drains into the New River, on the other side of which the major center of San Estevan (Bullard 1965) is located. Hammond (1973) mapped two outlying residential clusters to the south of San Estevan, and recent survey in this area has yielded evidence of nearly continuous, dispersed settlement east of the New River (Laura J. Levi, personal communication, 1990).

Major settlement boundaries between Pulltrouser Swamp and surrounding sites are formed by intervening swamps and rivers. Most likely, the dwellings dispersed around the arms of Pulltrouser Swamp represent outlying settlement that was affiliated (in terms of taxation, labor obligations, or both) with either Nohmul or San Estevan and do not represent politically discrete units of settlement. Thus, with largely an agrarian role in the settlement system of northern Belize, the sites bordering Pulltrouser Swamp may inform us about patterns of prehistoric agriculture and use of interresidential space in a way that data from highly maintained and frequently renovated "central places" will never be able to do.

A Model of Agricultural Land Use

Results of 1979 and 1981 settlement mapping efforts indicate that residential platforms are dispersed around the southern and easternmost arms of the swamp. Preclassic settlement is concentrated on the higher terrain where the soils are superior (i.e., at K'axob, Kokeal, and Yo Tumben; Fry 1983; Harrison 1983; McAnany n.d.). Expanded excavation at K'axob, in particular, has yielded evidence of a thriving Middle Preclassic village settlement (McAnany n.d.) while the testing program at Tibaat has indicated the predominantly Early Classic genesis of the settlement located between the two northern arms of the swamp (Walling and Minc n.d.). In general, the extent of settlement indicated in figure 8-4 represents cumulative settlement growth through the Terminal Classic. Approximately half of the 348 platforms at Pulltrouser Swamp are grouped around a central courtyard or patio, suggesting a strong corporate or extended family settlement structure (McAnany 1986). Distance between platform complexes within each settlement cluster ranges from 50 to 100 meters. A program of test excavations conducted between platforms yielded limited evidence of ancillary or residential ground-level structures (McQuarie n.d.). Nevertheless, large plots of land existed between structures that could have been utilized as agricultural plots.

The agricultural history of this area, as reconstructed through palynology, macrobotanical and phytolith analysis, and stratigraphic interpretation of agricultural features (raised fields and canals), involved a sequential shift from cultivation in the uplands to the west of Pulltrouser Swamp to construction of raised fields in the eastern and southern arms of Pulltrouser Swamp (Turner 1983c; Miksicek 1983; Wiseman 1983). In the topographically higher areas to the west of Pulltrouser Swamp, pedogenic processes have acted upon the limestone substrate to produce clay-based soils that are moderately fertile. As sandy, gravelly, silty, or loamy clays, these soils tend to occur on the tops of low, broad ridges and, as Hammond (1974:177) has noted, are

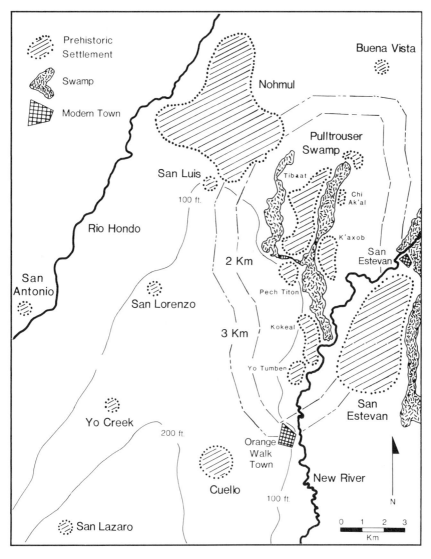

Figure 8-4. Settlement Bordering Pulltrouser Swamp and Surrounding Area with Two- to Three-Kilometer Infield Zone Delimited

invariably the locations of ancient Maya settlement. On the basis of pollen analysis from Pulltrouser Swamp, Wiseman (1983:114) has suggested that the dearth of arboreal pollen in the lower levels of raised-field fill may indicate that these fertile upland zones were already denuded of trees and extensively cultivated at the time of raised-field construction. The demography of modern Belize lends support to this suggestion. Today, colonizing populations in sparsely settled Belize are radiating into exactly these upland zones for *milpa* farming and ranching as well as the development of sugarcane and pineapple plantations (Wright et al. 1959).

If these uplands (area over 100 feet in elevation on figure 8-4) were completely denuded and brought under cultivation by A.D. 100 (a conservative estimate), then what are the implications for further agricultural expansion at a time when regional population levels were rising and local political economies expanding? In order to frame the soil and pollen data in a settlement and land-use perspective, the Preclassic Pulltrouser Swamp regional landscape can be modeled as partitioned into infield and outfield components. In such a model of agricultural land use, spaces close to the residence contain fields that are repeatedly tilled rather intensively (infields). In addition, outfields are cultivated in a more extensive, shifting fashion. Infields are generally defined as cultivation plots located within a forty-minute walk (two to three kilometers) from the residence (Killion 1987a; Wilk 1983). According to this perspective, much of the uplands west of Pulltrouser Swamp could be classified as infield land with areas outside of this radius classified as outfields. Cultivation of distant outfield areas often involves the establishment of an ancillary residence or field hut, which may be used for sleeping, eating, refurbishing agricultural tools, crop storage, and as a work area for processing harvested crops previous to transport back to the main residence (Killion 1990). Areas of upland outfields could have been used by several communities located in the region of Pulltrouser Swamp and, at present, no archaeological method exists for linking an agricultural settlement to a particular outfield location (see Santley, this volume, for a provocative argument regarding the identification of "near" outfield locations in the formative lowland setting of Matacapan).

The two- to three-kilometer infield boundary surrounding Pulltrouser Swamp is depicted graphically in figure 8-4. This zone circumscribes the three arms of the swamp and the modern town of Orange Walk and assuredly overlaps with the field zones of the prehistoric communities of Nohmul, San Luis, San Lorenzo, Cuello, San Estevan, and Buena Vista. The degree of overlap is underrepresented for two reasons: (1) the extent of settlement is not known for many of the surrounding settlements, and (2) the differential fertility of soils within the infield zone has not been taken into account. Nevertheless, delineation of this infield zone around Pulltrouser Swamp emphasizes the rather tight settlement packing of this area. In fact, this spatial application of the infield-outfield model indicates that, at least by the Classic period, outfields were probably not a component of the agricultural system, suggesting a nearly complete reliance on fields within three kilometers of settlement. In this light, the reclamation of the wetlands represents a rather predictable expansion of the infield zone to include the otherwise nonarable swamp. The period of time during which this wetland reclamation began has been notoriously difficult to identify due to the lack of sealed deposits. A suite of radiocarbon dates and the inclusion of diagnostic ceramics, however, bracket the Late Preclassic and Early Classic as periods during which raised-field construction was initiated (Turner and Harrison 1983:248).

From multiple (if admittedly thin) strands of evidence, therefore, we can propose a model of the agricultural history of this area as one of initial clearance and cultivation of the high ground within and around Pulltrouser Swamp employing an infield-outfield agricultural strategy. Regional settlement data coupled with local paleoenvironmental data suggest that during the Late Preclassic or Early Classic period a retraction of the agricultural system and a concomitant intensified focus on "settlement agriculture" occurred (Killion, this volume)—a specialized component of which was swamp reclamation. This reclamation involved not only the construction of raised fields and canals but, if the ethnographic record is any indication, also signaled an intensified focus on plot maintenance tasks such as mulching, weeding, and hoeing.

Stone Tools: Types and Contexts

Lithic collections (stone tools and debitage) from Pulltrouser Swamp are the result of seven different programs of artifact recovery: (1) a 10 percent random sample of all platforms with test excavations (1.5-meter squares) in both the construction fill and the adjacent middens, (2) judgmental test excavations, (3) excavation of a sample of surface features and subterranean chambers, (4) horizontal excavation of two platforms and one ground-level structure, (5) test excavations near and between platforms (interresidential space), (6) raised-field excavation, and (7) surface collections.

The most ubiquitous tool form recovered from these excavations was the oval biface (fig. 8-5A–C). It is present in all temporal contexts from the Middle Preclassic to the Terminal Classic (McAnany 1986). These teardrop-shaped tools were manufactured from high-quality northern Belizean chert, which outcrops about thirty kilometers to the south where the site of Colha is located. Large nodules, up to one meter in diameter, were quarried at Colha. In the region of Pulltrouser Swamp, in contrast, only large cobbles of brittle chalcedony that is unsuitable to the fabrication of large bifaces were available. The absence of technological indicators of chert tool production in the debitage from Pulltrouser Swamp (McAnany 1986, 1989; Shafer 1983), coupled with the abundance of tool production debris in concentrated debitage deposits throughout the settlement of Colha (Shafer and Hester 1983), suggests that from the Late Preclassic to the Terminal Classic (300 B.C. to A.D. 1000) finished tools were moved from Colha to the settlements of Pulltrouser Swamp, presumably in the social context of an exchange relationship.

In their pristine state, oval bifaces average 138 millimeters in length, 60 millimeters in width, and 20 millimeters in thickness (fig. 8-5A). Although bifaces are common throughout the Maya lowlands, the slenderness of the oval biface separates it from other thicker celt and tranchet-bit forms (Shafer 1983), which were probably used primarily for forest clearance. One method of hafting the oval biface has been indicated by the fortuitous discovery of a hafted water-logged specimen in a raised-field canal along the Rio Hondo (Puleston 1976). The

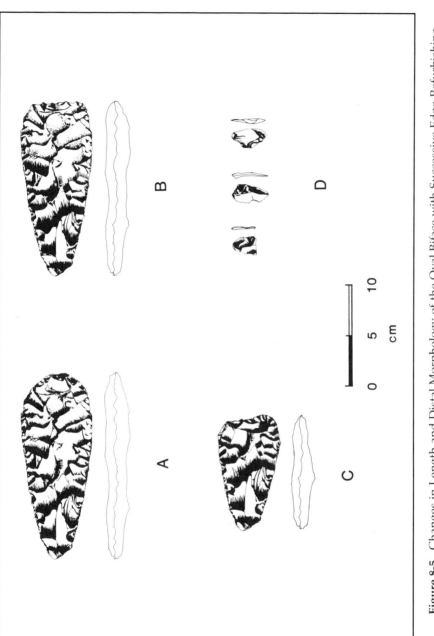

Figure 8-5. Changes in Length and Distal Morphology of the Oval Biface with Successive Edge-Refurbishing Events (A–C) and Types of Debitage Produced by Edge Maintenance (D)

narrow end of the tool has been inserted into a simple socket haft at a right angle to the blade in a manner similar to a modern steel ax (this specimen is illustrated in Shafer and Hester 1986); this tool had either been lost or, more likely, placed in the water-logged canal context in order to tighten the haft. Other methods of hafting the oval biface are possible, including a right-angled hoe haft. The majority of oval bifaces recovered from settlements and raised fields of Pulltrouser Swamp bore evidence of distal edge refurbishing or were broken fragments. As a hafted tool, the oval biface was repeatedly shortened by the detachment of flakes primarily from the dulled working edge, giving the resharpened tool a squared distal morphology (fig. 8-5B and C). Debitage resulting from biface resharpening was commonly recovered from excavations at Pulltrouser Swamp. Attribute analysis of this debris (illustrated in fig. 8-5D) indicates the presence of characteristics that are diagnostic of biface resharpening: that is, the presence of edge damage (rounding) and use polish (discussed below) on the platform and dorsal faces and the presence of obtuse angled platform-to-ventral surfaces. The excavated context of these flakes is informative of the place at which tool refurbishing took place. While material may be transported through sweeping or general site maintenance, the small size of these flakes (generally less than one centimeter in length) renders it unlikely that they would be transported far beyond their location of discard (Hull 1987).

Examination of edge damage and use polish on the flakes detached from the oval bifaces (McAnany 1986, 1989; Shafer 1983:figs. 12-6, 12-7) indicates that most edges were rejuvenated in order to maintain a sharp edge. That is, many of the flakes that could be identified technologically as having been detached from a biface bore platforms that had been rounded and polished through repeated contact with an abrasive, nonpercussive, high-silicate medium. The high clay content of the soils around and in Pulltrouser Swamp fits this description exactly. Furthermore, the inclusion of water lily pollen (Wiseman 1983:110–13) as well as macrobotanical material such as maize husks and charcoal (Miksicek 1983:104) in soil samples from the raised fields themselves suggests that these plots were indeed mulched. Such mulching would further enhance the phytolith content and abrasive polishing capabilities of the contact material (see Shafer 1983 for a discussion of the identification of this polish). In short, patterns of wear suggest that the oval biface was an agricultural tool; as such, it was probably used for field maintenance tasks (weeding and tilling) in a manner similar to the ethnographic accounts cited above. In figure 8-6 three common patterns of oval biface breakage are illustrated and the resulting proximal, medial, and distal fragments identified. The upper breaks on figures 8-6A and 8-6C correspond to breaks within the haft, while the lower breaks on figures 8-6A and 8-6B are snap breaks that occurred at the base of the haft. Snap fractures or bend-breaks that leave a diagnostic lip (fig. 8-6B) are the most common form of breakage at Pulltrouser Swamp. This type of breakage generally occurs when a stone tool is used in a twisting or prying motion and the physical resistance of the contact material exceeds the tensile strength of the stone. Such breakage is not surprising given the evidence—both from ethnographic sources (summarized above) and from use polish on the tools themselves—that the working end of these tools often penetrated the soil.

Tool breakage obviously occurred at the place of use. For agricultural tools, this fact translates into breakage at the location of agricultural plots. Excavation in midden deposits adjacent to residential platforms, however, indicates that selective transport of broken tool fragments back to the residence occurred. Proximal fragments, still in the haft, may have been returned to the residence only to be discarded when the haft was refitted or retooled with a new blade. Based on ethnographic data, Keeley (1982:800) has noted that hafting is generally more time consuming than the manufacture of the stone tool itself; therefore, hafts are more likely to be reused even after the tool blade has been broken. Binford (1979:263) likewise has noted that broken tools are often transported back to the residence where they are repaired and the useless portions are discarded. Thus, the recovery of proximal oval biface fragments in a residential context would seem to be indicative of the place at which retooling occurred rather than the location of tool breakage.

Medial fragments are also commonly found in domestic middens. These fragments were frequently recycled as cores or metate-pecking stones (McAnany 1988) and thus may have been transported back to the residence because they represented a large mass of recyclable stone in a stone-impoverished area. Distal or broken bit fragments, however,

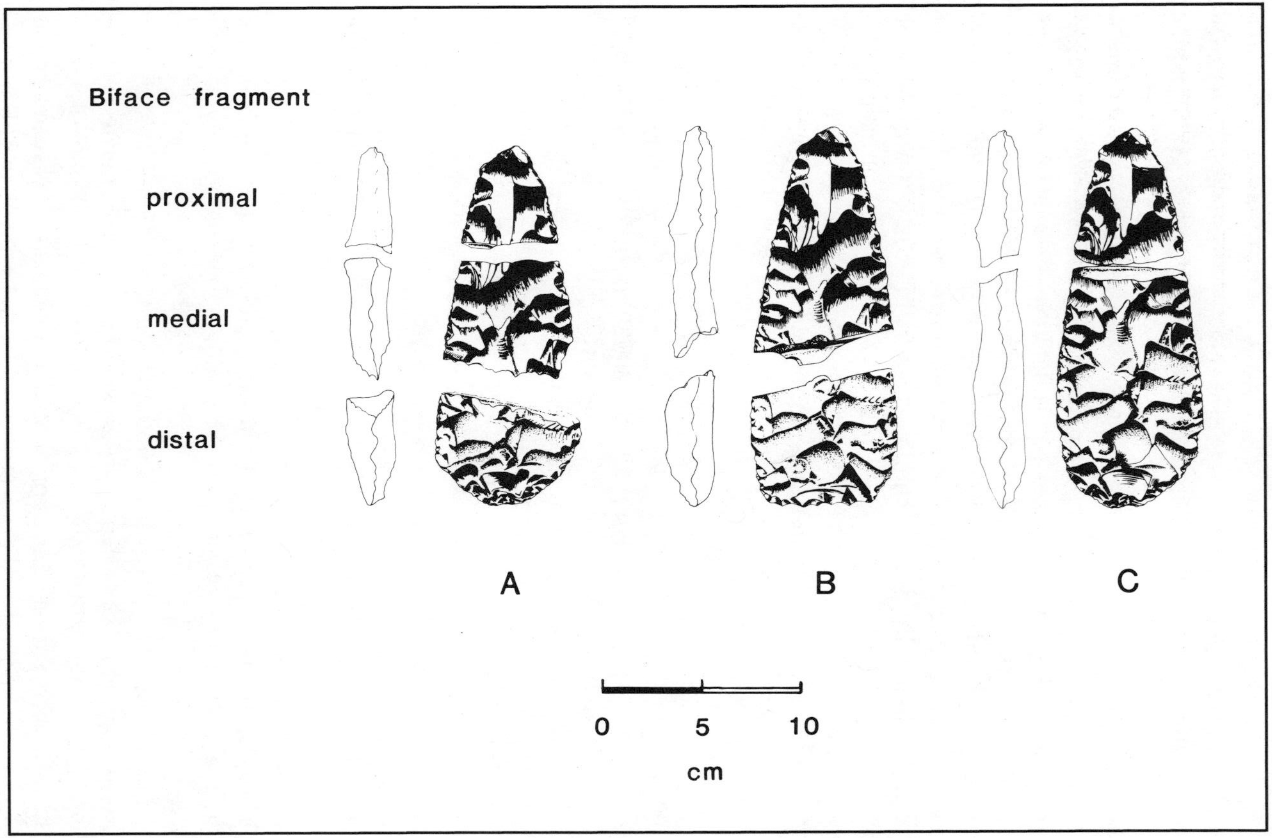

Figure 8-6. Patterns of Oval Biface Breakage (A–C) with Resultant Fragments Identified

have limited potential for recycling due to their small size and because they tend to be riddled with internal percussion fractures. Distal fragments, theoretically, are likely to be discarded at the location where tool breakage occurred, possibly at field locations, and to undergo no further transport. In fact, these tool fragments are underrepresented in the domestic middens except for some interesting exceptions that are discussed below.

These characteristics of the oval biface—refurbishing (including edge resharpening and retooling) and transport of broken fragments—suggest that these tools were highly curated (McAnany 1988). The presence of diagnostic tool fragments and debitage in residential middens indicates that these tools were transported between the place at which they were used—the cultivation plot—and the place at which they were refurbished—the residence. As such, we can expect that changes in the locational patterning of tool fragments as well as debitage will be informative of changes in agricultural practices.

STONE TOOL REFURBISHING AND DISCARD LOCATIONS

In order to examine locational and temporal patterning in broken tool fragments and refurbishing debris, the lithics from platform midden contexts and interplatform contexts (both surface and excavated) were tabulated separately. Within each context lithics associated with temporally diagnostic ceramics were separated by temporal period. Temporal designations follow those of Fry (1983:197) and are printed on table 8-1. As with most archaeological data, a problem exists with small sample sizes for the earlier periods under study. Consequently, the later periods (Late and Terminal Classic) tend to contribute disproportionately to the sample.

Biface Tool Fragments

Oval biface fragments (proximal [haft end] and distal [bit end]) from platform-associated midden and interplatform (interresidential) contexts are compared by temporal period in table 8-1. Within each context

Table 8-1. Biface Tool Fragments from Residential Middens and Interplatform Contexts

	Platform Midden Oval Biface Fragments							Interplatform Oval Biface Fragments					
	Proximal		Distal		Total		Proximal		Distal		Total		
Temporal Period	N	(e)*	N	(e)	N		N	(e)	N	(e)	N		
Early and Middle Preclassic (1200–400 B.C.)	3	(2.2)	1	(1.8)	4		3	(2.1)	1	(1.9)	4		
Later Preclassic (400 B.C.–B.C./A.D.)	2	(1.1)	0	(0.9)	2		0	–	0	–	0		
Proto- to Early Classic (B.C./A.D.–A.D. 400)	9	(7.7)	5	(6.3)	14		6	(5.3)	4	(4.7)	10		
Middle Classic (A.D. 400–600)	9	(6.6)	3	(5.4)	12		3	(2.1)	1	(1.9)	4		
Late Classic (A.D. 600–800)	9	(8.2)	6	(6.8)	15		5	(4.8)	4	(4.8)	9		
Terminal Classic (A.D. 800–1000)	36	(42.2)	41	(34.8)	77		50	(52.6)	49	(46.4)	99		
Total	68		56		124		67		59		126		
Chi-square value			6.972						2.021				
Degrees of freedom			5						4				
Probability value			0.223						0.732				

* (e) is the expected cell chi-square value; values less than 5, and there are several on this table, weaken the validity of the chi-square statistic and the significance of the probability value.

a chi-square statistic was calculated in order to test the assumption that no significant difference exists between the proportion of proximal to distal fragments. Expected cell chi-square values are printed in parentheses next to the frequency values, and the overall chi-square statistic, degrees of freedom, and probability values are printed at the base of the table. It should be noted that many of the cell frequencies are very small and have expected cell chi-square values of five or less. Thus, the results of this chi-square analysis should not be construed as statistically significant but rather illustrative of a general trend that may be rendered significant given a larger sample size.

In the residential middens distal fragments are underrepresented during the Preclassic period, although the sample size is very small. Still this pattern is the expected one if these tools were being used and broken in cultivation plots away from the residence (infields or outfields) and the hafts (with the still imbedded proximal end) were returned to the residence to be retooled. Throughout the Classic period, during which the nearby raised fields were constructed and interresidential space may have been converted to increasingly intensive agricultural uses, a steady increase occurs in the proportion of distal fragments in the residential midden peaking in the Terminal Classic. This period is the only one during which the frequency of broken distal fragments (41) exceeds the chi-square expected value of 34.8. The more or less equal ratio of proximal to distal fragments during the Terminal Classic suggests that either less retooling took place at the residence (and therefore fewer broken proximal fragments were discarded) or more agricultural activity took place close to the residence that resulted in a higher incidence of broken bit fragments discarded in residential middens possibly in an attempt to maintain stone-free cultivation plots.

The latter interpretation is supported by the identification of a similar pattern in the interplatform areas. In this context, the proportion of distal fragments also increases through the Classic until the Terminal Classic at which point it is on a par with proximal fragments. The small sample size of the earlier periods produces a statistically insignificant chi-square value for both contexts. The implications of the increasing proportion of broken distal tool bits at close proximity to the residences remain, however, and can be further explored by examination of biface edge-refurbishing debris.

Debitage from Biface Edge Refurbishing

Agricultural tools can be resharpened in the field, as they are dulled from use, or at the residence in anticipation of use. As mentioned above, when outfield plots located at some distance from the residence

Table 8-2. Debitage from Platform Middens and Interplatform Contexts

Temporal Period	Platform Midden Type of Debitage					Total N	Interplatform Type of Debitage				Total N
	Bifacial			Other			Bifacial		Other		
	N	(e)*		N	(e)		N	(e)	N	(e)	
Early and Middle Preclassic	6	(8.5)		19	(16.5)	25	1	(5.3)	14	(9.7)	15
Later Preclassic	10	(18.7)		45	(36.3)	55	6	(4.2)	6	(7.8)	12
Proto- to Early Classic	31	(35.7)		74	(69.3)	105	33	(31.5)	56	(57.5)	89
Middle Classic	17	(15.3)		28	(29.7)	45	0	(0.4)	1	(0.6)	1
Late Classic	33	(32.0)		61	(62.0)	94	16	(19.5)	39	(35.5)	55
Terminal Classic	144	(130.9)		241	(254.1)	385	164	(159.2)	286	(290.8)	450
Total	241			468		709	220		402		622
Chi-square value				10.509					8.367		
Degrees of freedom				5					5		
Probability value				0.062					0.137		

* *(e)* is the expected cell chi-square value.

are worked from a field hut or temporary shelter, then it is likely that tools will undergo refurbishing at that location. As a consequence, less edge-refurbishing debris will be found in the middens of the primary residence. In table 8-2 the ratio of flakes with bifacial platforms (debitage from biface refurbishing) is compared with nonbiface refurbishing flakes (flakes lacking bifacial platforms). As would be expected in a situation of increasingly intensive agricultural use of near-fields and interresidential spaces, the pattern of table 8-2 is one of a general proportional increase in biface edge-maintenance flakes relative to other types of flakes in the residential middens over time. The proportion of refurbishing flakes to other types of flakes is greatest during the Middle, Late, and Terminal Classic periods.

In the interplatform contexts the pattern is generally a similar one. Once again, a small sample for the Middle and Late Preclassic periods may be due to the less intensive use of these areas during the Preclassic. These patterns in the lithic data do conform to the expectation of increasingly intensive use of interresidential space as suggested by the land-use model. An expanded sample size encompassing a larger corpus of material from earlier periods and more interplatform excavations will be necessary for a critical and statistically robust test of the model.

DISCUSSION AND CONCLUSIONS

Examination of biface fragments and refurbishing debris has revealed two temporal trends: (1) an increase over time in the proportion of broken distal tool fragments relative to the hafted (proximal) end in both the residential middens and the interresidential spaces, and (2) an increase over time in the proportion of biface refurbishing debitage also in both contexts. In other words, patterned lithic remains from agricultural implements indicate that, through time, and in tandem with raised-field construction, oval bifaces were increasingly refurbished as well as broken in close proximity to the residence.

The oval biface itself has been argued to have been an agricultural implement utilized for the types of field maintenance tasks that are associated with increased cropping of a plot, specifically weeding and tilling the soil. Likewise, the increased visibility of debris from maintenance and breakage of these tools during the Classic period is argued to be a general indicator of increased emphasis on the cultivation of near-residential plots, both the raised fields and the plots located between residences. To what extent these patterns may be indicative of the attenuation of outfield plots cannot be determined by this analysis; however, shrinkage of the catchment area available to agriculturalists living near Pulltrouser Swamp is certainly indicated by recent settlement survey in adjacent zones.

The notion that architecturally vacant spaces between Maya platforms may have been used for plant cultivation is certainly not new (Thompson 1966); however, the concept that in some places these spaces were the *primary* field space of the Maya sheds an entirely new meaning and nuance of intensity on the term "garden city." Based on contemporary Ibo agriculture and settlement density, Netting (1977) several years ago suggested that during the Classic period spaces between Maya residences may have been utilized as permanent field plots. The pattern of residences dispersed across demarcated terraces in the Rio Bec region (Turner 1983a) plus the walled compounds at Coba (Folan, Kintz, and Fletcher 1983) and field walls of Cozumel (Freidel and Sabloff 1984:84–90) further indicate that some sort of interresidential field component was vital to prehistoric Maya agriculture. Santley, Killion, and Lycett (1986) have argued for the importance of infield plots, particularly during the final epoch of the Classic period when population levels peaked, and Killion and others (1989) have recently applied the model of interresidential agriculture to Sayil in northern Yucatan. Even during historic times when Maya population levels were relatively low, the important role and ubiquitous presence of "kitchen gardens" were noted by de Landa (Tozzer 1941), Roys (1931), Wauchope (1938), and Wilken (1971; see Turner, this volume, for clarification of the distinction between infield and kitchen garden).

The fact that the apparent population peak in this region occurred during the Terminal Classic while evidence of intensified agricultural use of near-residential space and the initiation of raised-field construction can be found in earlier time periods suggests that this process was not simply a response to pressure on the land. Political centralization (and its attendant costs) and increased participation in regional economic exchange networks (McAnany and Isaac 1989) are variables that cannot be ignored in terms of their general effects on agricultural production.

Finally, this case study from Pulltrouser Swamp has demonstrated that the dynamics of Precolumbian agricultural systems are not predicated solely upon the fabrication of a new tool form or upon amassing the labor necessary to construct agricultural features and thus reclaim areas marginal to agricultural production. Other long-term accretional effects of agricultural change often can be monitored with the archaeological record. In this case, increased agricultural maintenance tasks have been linked to distinctive patterns of tool wear and breakage; likewise, changes in the agricultural use of space between and near residences has been linked to changes in the location of discarded tool fragments and edge-refurbishing debris. This approach to prehistoric agriculture moves beyond the purely typological/evolutionary approach and, as such, can yield increased archaeological resolution of prehistoric agrarian systems. This approach also requires lithic analyses that are technological rather than purely typological and excavation strategies that sample architectural as well as nonarchitectural spaces within a settlement. As one of the symposium discussants remarked, "It's a hard row to hoe." But the harvest we reap will be worth it.

ACKNOWLEDGMENTS

The first draft of this chapter was prepared in 1987 while I was holding a Charles Phelps Taft Postdoctoral Fellowship at the University of Cincinnati. Since that time I have benefited from a cascade of feedback on this study ranging from helpful criticisms to blatant objections. My research on agriculture and stone tools has without a doubt been strengthened by these interactions and I wish to acknowledge and express my gratitude to the following individuals in particular: B. L. Turner II, William T. Sanders, Thomas W. Killion, Rhoda H. Halperin, and the two anonymous reviewers.

PART III

Prehistoric Cultivation, Landscape Modification, and Chemical Characterization

CHAPTER 9　　　　　　　　　　　　　　　　　　　CHRISTIAN J. ZIER

Intensive Raised-Field Agriculture in a Posteruption Environment, El Salvador

THE GOALS of this chapter are twofold. It is intended primarily as a description of uniquely well-preserved agricultural features in a volcanic environment. Secondarily, the chapter attempts to draw inferences about primitive agricultural practices in El Salvador in the context of a population recovering demographically from a major volcanic disaster.

Archaeological discoveries over the past quarter-century in the New World suggest that various forms of raised-field cultivation were practiced prehistorically in South, Central, and North America (Parsons and Denevan 1967; Harrison and Turner 1978; Flannery 1982; Fowler 1969; see review in Denevan 1982). Prehistoric raised-field agricultural features are reported from central and southern Mexico and adjacent portions of Central America and occur in both highland and lowland settings (e.g., Siemens and Puleston 1972; Siemens 1982; Seele 1973; Earnest 1976; Zier 1980). These features vary greatly in form and context and perhaps also function.

Raised fields are created either by mounding or ridging of soils. The discussion here will emphasize the latter form, although certain properties may be shared by both forms. Two very different types of function are associated with most raised fields: nonhydraulic and drainage (Denevan and Turner 1974:26–29). A third type of function, moisture retention, is probably facilitated to some degree by all raised fields as a result of increasing the surface area through which rainwater may percolate with a concomitant reduction of evaporation due to concentration of soil. However, only in extreme cases such as construction of *chinampas* (floating gardens) is moisture retention a primary function of raised-field construction.

As described below, the Cerén site is essentially a snapshot, frozen in time and place by the effects of a sudden, violent volcanic eruption. While agricultural features in place at the time of the eruption are clearly identifiable, little is known of the range of field and feature configurations that might have been used by the inhabitants of the site or the evolution in form that a particular field may have undergone during the course of a planting season. Thus, it cannot be stated with certainty that fields always assumed a ridged form or that ridging was not simply a seasonal practice.

Nonhydraulic functions provide the best justification for the creation of ridges and furrows described in this chapter. Nonhydraulic functions are potentially numerous and include increasing of fertility through mounding and concentration of humic topsoil or aeration of soil, facilitating weeding and harvesting, control of topsoil erosion, and maintenance of soil temperatures. Ridging of soil in such contexts may to a large extent be characterized as slope management through modification of surface contours. Raised fields established for drainage purposes tend to occur on a far grander scale than those of a nonhydraulic nature in terms of sheer volumes of earth displaced, and they do not appear functionally related to features described here.

On a very general level, Denevan and Turner (1974) note several traits that tend to characterize all raised-field agriculture. First, it is a technique that is employed in areas of low fertility and is usually a form of reclamation of marginal agricultural lands. Second, it tends to be used in cultivation of root crops although many exceptions exist. Finally, raised-field agriculture is frequently associated with relatively dense rural populations in the tropics.

THE CERÉN SITE AND
REGIONAL VOLCANIC ACTIVITY

The Cerén site was partially excavated by the University of Colorado in 1978 and again in 1989. The earlier excavations provided most of the data that are the basis for this chapter, although some new information did come to light in 1989. Extensive survey and excavations were undertaken in the Zapotitán Valley of west-central El Salvador for the purpose of studying the nature of human reoccupation of the region in the wake of a massive volcanic eruption (Sheets 1971, 1976, 1983). Widespread areas of central and western El Salvador were abandoned after the ca. A.D. 260 eruption of Ilopango Volcano. This volcano, presently a large water-filled caldera, lies sixteen kilometers east of the contemporary city of San Salvador. The Zapotitán Valley is centered forty-five kilometers west-northwest of Ilopango and nestles between two other major volcanic complexes, Santa Ana to the west-northwest and San Salvador to the east. The Balsam Range encloses the valley on the south (fig. 9-1).

Deposits of air-fall tephra and ash flows from Ilopango buried the Zapotitán Valley to an average depth of two meters (Hart and Steen-McIntyre 1983). Deposits nearer the source are considerably deeper. The ubiquitous white tephra, or ash, from the eruption is referred to locally as *tierra blanca,* or literally, 'white earth'. Field mapping of Ilopango *tierra blanca* exposures at various localities in central and western El Salvador (Hart and Steen-McIntyre 1983:22) has indicated that an area within a radius of approximately thirty kilometers surrounding Ilopango was blanketed with a *minimum* of one meter of tephra and that areas seventy kilometers from the source received at least fifty centimeters of deposit. Original tephra deposits were probably greater than those recorded at most localities but have been diminished through erosion and compaction. At the large archaeological site of Chalchuapa near the Guatemala border seventy-five kilometers west-northwest of Ilopango, *tierra blanca* deposits are thirty centimeters deep. While *tierra blanca* is known chiefly from El Salvador, possible Ilopango tephra has been identified in lake cores from the central Peten in Guatemala (Sheets 1983:5–6). It is estimated that an area of perhaps three million square kilometers, stretching northwestward from west-

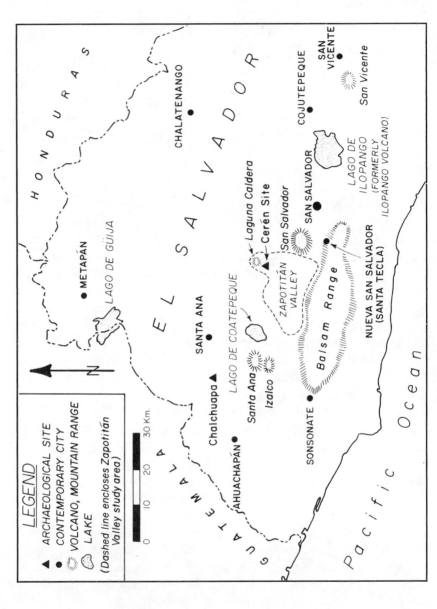

Figure 9-1. Map of Central and Western El Salvador Showing Study Area and Locations of Major

ern Nicaragua and Honduras to northern Mexico and southern Texas, was affected to some degree by the eruption (Hart and Steen-McIntyre 1983:28–30).

Although most Preclassic sites predating the Ilopango eruption remain buried, archaeological evidence from the Zapotitán Valley and elsewhere (Sheets 1976, 1983; Sharer 1974, 1978) suggests that El Salvador was heavily populated immediately prior to the eruption. The eruption would probably not have proved fatal for most of the valley's inhabitants but would certainly have forced immediate abandonment. It is believed that the valley lay vacant for approximately three centuries before occupation by agriculturalists was renewed.

The Cerén site lies along the northwestern margin of the valley near the Rio Sucio, the valley's major drainage. It dates to A.D. 590 ± 90 and is probably representative of the earliest phase of human reoccupation. The Cerén site is built directly atop *tierra blanca* from the great Ilopango eruption. The site was in turn buried beneath approximately five meters of tephra from the localized eruption of a small volcano known as Laguna Caldera. The site was inhabited at the time of the Laguna Caldera eruption. Beneath the deep tephra is a well-preserved record of Early Classic agriculture that has yielded data about crop species, seasonality, planting practices, and the spatial interrelationships among agricultural fields and residential units.

Excavations have revealed the remains of six structures in three compounds or complexes (Zier 1983; Sheets and McKee 1989). The structures are interpreted to be nonelite residences and various associated outbuildings. Portions of an agricultural field were exposed in two large test pits a few meters from one of these structure complexes. Two structures in one complex were initially sectioned and thus exposed during a land-leveling project undertaken by the Salvadoran government. The residence (Structure 1) (fig. 9-2) exhibits remarkably well-preserved standing wattle-and-daub walls and columns, packed clay floors, and in situ archaeological evidence of numerous discrete activities including ceramic manufacture, weaving, and food storage. The artificial section through Structure 1 exposed several superimposed floors underlying the latest one. Burned palm thatch on floors indicates the general manner in which the structures had been roofed. Outbuildings consist of bare, slightly elevated platforms with major post supports along the margins (Zier 1983; Sheets and McKee 1989). Distances among structure complexes range from thirty-five to fifty-five meters.

The deeply cut profile through Structures 1 and 1a, which extended southeastward from the house approximately thirty meters to the edge of a cut bank overlooking the Rio Sucio, clearly displayed the volcanic history of the locality (fig. 9-2). The basal unit is a reddish brown, well-developed clayey soil of Preclassic age, formed on earlier tephra. This buried soil is ubiquitous in the Zapotitán Valley, where it is commonly exposed in deep road cuts. It was an excellent soil for plant growth (Olson 1983b:52–53) and supported a Preclassic Maya agricultural population.

Immediately overlying the Preclassic clay is the Ilopango *tierra blanca*. The thickness of this deposit at the Cerén site is just twenty-five to forty centimeters, considerably thinner than the Zapotitán Valley average of about two meters. This fact indicates that the Cerén site occupied a relatively unstable erosional surface. Plainly exposed in the Cerén site profile are pits that had been excavated through the *tierra blanca* and into the underlying clay, presumably to extract raw material for house construction or other purposes.

The massive Laguna Caldera deposit, known as the Cerén tephra, directly overlies the *tierra blanca* tephra—and the Cerén site. This deposit exhibits coarsely and finely bedded tephra units that include, in the lower levels, large scoria bombs (Hart 1983). The thickness of the Cerén tephra is approximately five meters.

AGRICULTURAL FEATURES OF THE CERÉN SITE

The test pits in which portions of the agricultural field were exposed are seventeen meters apart and lie generally to the south and west of the Structure 1 complex. Further excavation, such as clearing of intervening areas between the units, was not attempted because of the considerable depth of overburden and the sudden onset of the 1978 rainy season. Test Pit 2 was expanded somewhat in 1989 following an eleven-year hiatus, and an additional small area of the original ground surface was exposed (Murphy 1989).

Each test pit revealed sets of parallel cultivation ridges with inter-

Figure 9-2. View Southwest toward Artificial Cut Exposing a Cerén Site House (Structure 1) and Section of Local Stratigraphy. At base of exposure is dark Preclassic clay (A) formed in ancient volcanic materials. Overlying the clay and directly underlying the house is light-colored *tierra blanca* tephra (B), eroded at this locality to a thickness of just twenty-five to forty centimeters. Superimposed house floors (C) are visible in profile beneath the latest floor. Directly overlying the house is the massive deposit of Cerén tephra (D) from Laguna Caldera that is five meters thick at this location.

mediate troughs or furrows (Zier 1980:68–71). The first pit exhibited six ridges ten centimeters high, spaced at regular intervals of thirty-eight to forty-two centimeters. Ridge crests were flattened and irregular. Drainage was into a trough that crosscut the six parallel rows at approximately a right angle. Slope along the trough ranged from 0.5 to 4.5 degrees. The prehistoric surface was littered with casts of plants.

In the second pit (fig. 9-3) were found four parallel rows with ridge crests fifteen to twenty centimeters high. The orientation of rows here was approximately the same as that of the crosscutting row in the first pit, with drainage to the southwest. The slope of the trough in this pit was 1.0 to 2.0 degrees. Clusters of circular, vertical casts of maize stalks, ranging in diameter from six to thirty millimeters, were found at regular intervals of about forty-eight centimeters along the ridge crests. Surrounding most vertical cast clusters were small concentrations of leafy casts, pressed into the ground surface and occurring in the lowest few centimeters of tephra deposit (fig. 9-4). While only individual, isolated maize stalk impressions were identified in 1978 (Zier 1980, 1983), further investigation in 1989, which included casting of plant molds with dental stone (plaster), demonstrated that the stalks in fact occurred in clusters, generally of four (Murphy 1989).

IMPLICATIONS FOR AGRICULTURAL PRACTICES

The *tierra blanca* tephra that was farmed at Cerén constituted a poor agricultural soil. It exhibits minimal visible evidence of weathering and a paucity of organic matter. Numerous elements important to plant growth are poorly represented, including phosphorus, magnesium, potassium, calcium, and manganese. Nitrogen-bearing compounds are about average for tropical volcanic soils, as is soil pH (7.1) (Sheets 1982:106; Olson 1983a:58–59, 1983b). Agricultural practices in evidence at the site are both a reflection of and a response to those soil conditions.

The advantages that may accrue from creation of ridges have been discussed previously. The use of artificial ridges at Cerén probably served any of a number of purposes. It is noted above that the occupational surface at Cerén is an erosional surface from which a significant

Figure 9-3. Overhead View of Test Pit 2 with Exposed Portions of Cornfield. Ridges are clearly visible in profile at top of photo; small bits of flagging mark locations of young maize plants along ridge crests. (test pit dimensions: 2.5 by 2.5 meters; arrow points north)

Figure 9-4. Cast of Young Maize Leaf, Preserved in Fine-grained Cerén Tephra Overlying Cultivation Ridges in Test Pit 2 (length of impression: ca. three centimeters)

amount of *tierra blanca* tephra had already been naturally removed. The slope of the ground surface in the immediate vicinity of the house and field(s) is 0.5 to 4.0 degrees but is considerably greater throughout much of the general site surroundings. Clearly, the unstable *tierra blanca* was subject to degradation through sheet washing, and a planted crop was jeopardized by heavy or sudden rains. The use of crosscutting rows as evidenced in the two test pits served to reduce the rate of flow of rain runoff and thereby retard soil erosion. Presumably, excess moisture was directed into the nearby Rio Sucio.

Field ridging would have encouraged moisture retention during the periodic short, dry episodes that are characteristic of the Salvadoran rainy (growing) season. Moisture retention may have been a more pressing concern at Cerén than, for example, among the earlier Preclassic occupants of the area, who farmed a well-developed clayey soil. The poorly compacted, grainy *tierra blanca* at Cerén may have tended to desiccate quickly under dry weather conditions.

Ridging or mounding of soil around maize plants is sometimes practiced today in Central America during the latter part of the growing season. Protection of plants against wind damage is one justification for such activity. It is unlikely, however, that the Cerén agricultural features supported a late season crop, for reasons discussed below. Heaping of soils around mature plants may indeed have been practiced at Cerén, but archaeological evidence of such has not yet been uncovered.

Finally, creation of rows may have concentrated the limited organic nutrients into ridge crests where seeds were planted. Intentional concentration of organic materials is not viewed as a primary motivation for ridged-field construction at Cerén but may rather have been an ancillary benefit.

Other indicators exist of limited soil productivity at the site. It appears that the portion of the Cerén field exposed in Test Pit 1 was fallow at the time of the Laguna Caldera eruption. The surface in this pit looked trampled and neglected, and ridges were flattened and irregular. Plant debris on the surface was far more abundant than in the second pit, but only limited evidence was found of actual growing plants. No plants were found growing at intervals along ridge crests as in Test Pit 2. Leaf casts here were more brittle, suggesting that the plants were already dead when buried. Fallowing for one or more growing seasons is ordinarily a means of allowing exhausted or nutrient-poor soils to recover before reuse.

Pollen analysis of Cerén field soils was largely inconclusive but did yield one important result: identification of abundant small charcoal fragments (Short 1983). Introduction of burned organic material to the soil may reflect in-place burning of secondary growth in the field prior to planting. It may indicate further efforts to supplement a fundamentally poor soil through release of phosphates and nitrates.

One possibility that cannot be overlooked at Cerén is that the underlying rich Preclassic soil was tapped for crop growth. The mantle of *tierra blanca* in the Cerén field is just thirty-five to forty centimeters deep. The Cerén inhabitants were certainly aware of the presence and depth of the earlier clay as well as its general properties, as indicated by the clay mining pits near the house. Digging sticks such as those represented by several perforated stone disks (digging-stick weights) found in association with the Cerén structures were probably used to make holes for depositing seeds. Affixed to a wooden shaft, the stone weight could have been employed effectively to standardize the depth at which seeds were planted (fig. 9-5) (Zier 1980:73–74; Hummer 1983). Seeds may also have been planted in deep pits that were then refilled with soil, either immediately or gradually.

Contemporary highland and lowland Maya farmers plant corn seeds just a few centimeters below the surface (e.g., Redfield and Villa Rojas 1934:44; Vogt 1969:45–46), and indeed many strains of maize will not germinate effectively if the seeds are planted too deeply. However, Pueblo Indians of the American Southwest commonly plant at depths of fifteen to thirty centimeters and perhaps on occasion up to sixty centimeters (Cushing 1920:181; Titiev 1944:182–83; Weatherwax 1954:64; Bradfield 1971:5–6). In the case of the Pueblos, deep planting is practiced almost solely to take advantage of subsurface moisture. It seems probable, therefore, that deep root systems could have developed in the Cerén field that penetrated the buried Preclassic clay, particularly inasmuch as roots would have tended to respond to stimuli from the soil. What is not at all certain is the capability of this particular strain of maize to germinate at great depth.

At the very least, even if it was not used as a nutrient source for plant growth, the clayey Preclassic soil probably tended to trap and retain moisture at the contact with the overlying *tierra blanca.* The importance of the Preclassic soil to post–Ilopango eruption agriculture will not be known until additional work is undertaken in the Cerén vicinity.

Although water management is evident at the Cerén site, it is unlikely that irrigation was being practiced. The absence of microsorting and siltation in the Cerén furrows, in combination with the lack of a suitable permanent water source, dictate against the use of irrigation. Had the Rio Sucio been used as a source, a ditch 3.5 to 4.0 kilometers in length would have been required (Sheets 1982:107–08; Zier 1980:72–73).

Figure 9-5. Contemporary Salvadoran Farmers with Digging Sticks (Zapotitán Valley, 1978). It is common practice in highland Central America today to drop several maize kernels in a single hole during planting. Stone-weighted digging sticks may have been used at the Cerén site to enable placement of seeds at sufficient depth so that the nutrient-rich Preclassic clay underlying Ilopango *tierra blanca* could be tapped.

In the absence of irrigation, it is almost certain that the young plants in Test Pit 2 represent wet-season agriculture. Contemporary farmers in El Salvador plant corn at the beginning of the wet season, at about the time of the first rain. This event usually occurs in early to mid-May. Young plants with stalk diameters equivalent to those in the Cerén field can be expected within a few weeks of planting. It has therefore been estimated that the Laguna Caldera eruption occurred sometime in May or June and more likely the latter month (Zier 1980; Sheets 1989).

EARLY CLASSIC PERIOD AGRICULTURE IN A POSTERUPTION ENVIRONMENT

The Cerén site documents occupation of a marginal agricultural environment during the Early Classic period. The demographic implications of raised-field agriculture in nutrient-poor soils seem quite clear: that reoccupation of the region following the Ilopango disaster occurred at least in part as a response to population pressure elsewhere. The local depositional characteristics of the unweathered, nutrient-poor *tierra blanca* within the Zapotitán Valley at the time that it was reinhabited undoubtedly dictated the manner in which reoccupation took place.

It is probable that the first agriculturalists settled in the hilly areas of the valley margins, where water erosion had stripped off or greatly reduced the depth of the *tierra blanca* mantle. Erosion in these areas facilitated access to the older, preeruption soils and would also have led to relatively early regeneration of native vegetation. The northern and eastern valley margins, which include the Cerén site vicinity, were most suitable to early habitation because slopes are not very steep. Other areas, particularly the foothills of the Balsam Range on the south side of the valley, are extremely rugged. They are sparsely inhabited today and are suitable only for coffee production. Archaeological data suggest that the Balsam Range foothills were hunted prehistorically but not farmed (Black 1983).

The earliest incursions into the valley floor would have occurred after establishment of the valley margins and probably followed major watercourses such as the Rio Sucio and its tributaries. Riparian zones associated with these streams may have experienced erosional stripping of *tierra blanca* due to seasonal flooding. The remainder of the valley floor was subsequently reoccupied as the vast *tierra blanca* deposit weathered in situ into a productive agricultural soil. This soil eventually sustained a large Late Classic resident population (Black 1979, 1983; Zier 1981).

Excavation, testing, and remote sensing data from Cerén suggest a prehistoric pattern of spaced residential units with intermediate areas utilized for crop production. Using information about the Cerén site that was available at the time, Sheets (1982:113–14) developed estimates of plant productivity relative to available agricultural space and concluded that the quantities of maize grown in the Cerén fields would have been inadequate to sustain families occupying the houses for a year. New information about the site has since been unearthed. It is now known that clusters of four maize plants, rather than individual plants, were grown at Cerén; however, the amount of space available to each household is even less than the one-third hectare originally estimated by Sheets. His overall conclusions appear to be valid yet. Sheets (1982:114) notes, for example, that at Todos Santos in the Guatemalan highlands, a family of five requires about 1.2 hectares to produce maize sufficient in quantity to satisfy annual needs. Logically, supplementary food sources would have been required at Cerén. A pattern of regional settlement in which early occupation focused on the valley margins is again identified as adaptively advantageous, these areas affording access to the greatest range of nondomesticated food resources.

Only the use of seed crops is clearly documented at Cerén. In addition to maize plants found in situ in the field were small beans in two storage pots located within one residential structure (Zier 1983:131). Seeds of chile peppers and *uhushte,* a wild edible plant, also have been tentatively identified on the basis of recent work at Cerén (Sheets and McKee 1989). Historically, indigenous farmers in El Salvador have utilized an array of root and tree crops in addition to seed crops. Available root crops include *jicama,* manioc, and a form of wild onion; tree crops include *ramón, nance,* cacao, and avocado (Sheets 1982:116; Zier 1981:109–10; see Linares, Sheets, and Rosenthal 1975; Daugherty 1969; Smith 1980). Other edible plants occur in the area such as *Amaranthus* and *Chenopodium,* although data demonstrating ethnographic use are scanty. Finally, hunting, fishing, and shellfish collecting are known to have served to supply important dietary supplements in El Salvador until relatively recently (Sheets 1982:116).

It is suggested that the occupants of the Cerén site and other early resettlement communities employed an approach to agriculture that combined intensive farming of infield areas adjacent to residences with slash-and-burn farming of nearby hilly uplands (Zier 1981:111–12; Sheets 1982:114). Agriculture was in turn supplemented with collecting of wild plant foods and hunting, fishing, and shellfish collecting. (Remains of a bivalve and an edible snail were found during recent excavations; Sheets and McKee 1989). The uplands near the settlements provided the greatest hedge against starvation in the marginal agricultural environment of the time. Upland areas were probably the first to witness a return to preeruption faunal and floral conditions due to erosional loss of volcanic deposits, and as such they offered relatively stable and varied food resources at a time when valley-floor food production potential was low. The subsistence strategy was diversified and therefore adaptive. The Early Classic occupants of the posteruption Zapotitán Valley may have been much more the hunter-gatherers than their traditional sedentary reputation would lead us to believe.

CONCLUSIONS

The relationship between human settlement and land use in the post-Ilopango Zapotitán Valley is as much a matter of theory and inference as archaeological fact. The irony, of course, is that the type of volcanic activity that permits a detailed examination of one prehistoric locality, frozen in time and space, will also tend to obscure the larger system of which that single locality is a part. Given present information, it is not known if the Cerén site is representative of its time in terms of setting, intrasite spatial patterning, and agricultural practice.

Agricultural features preserved in the Cerén test pits suggest a short fallow system. Almost certainly inadequate to sustain the community using this system, it must have been supplemented to a great extent by agricultural and nonagricultural exploitation of more distant areas. That this subsistence strategy was largely effective is strongly suggested by the presence of multiple superimposed floor levels in the sectioned Cerén house (Structure 1), which indicate that the community had persisted in place for some time. The cultural and economic significance of the configuration of residences within the Cerén community is uncertain. While residential patterning is obviously related integrally to the intensive component of the agricultural system, full interpretation of community dynamics is not yet possible. It is beginning to look as if the various structure complexes at Cerén constitute components of a true village (Sheets and McKee 1989), possibly well organized spatially. Only when additional work is completed will we know the total size and crop composition of the Cerén field(s) and the number, nature, and overall distribution of dwelling units.

On a grander scale, one cannot help but be struck by the tremendous effect of volcanism on prehistoric inhabitants of the southern periphery of the Maya region or by the cultural resiliency that was displayed in the face of repeated natural disasters. Prehistoric settlement throughout much of the highland area of Central America was dictated largely by factors relating to volcanism: elevation, slope and drainage, soil fertility, and even to a certain extent climate. Unlike most areas of the prehistoric world, however, this landscape was dynamic—even volatile—in the short-term sense. The Cerén site represents both demographic readjustment and agricultural adaptability in such a context.

ACKNOWLEDGMENTS

Research described in this chapter was supported by National Science Foundation Grant No. BNS 77-13441 to the University of Colorado. Payson D. Sheets served as Principal Investigator and Project Director. The following project participants are thanked: Marilyn P. Beaudry, Kevin D. Black, Susan M. Chandler, William J. E. Hart, Jr., Richard P. Hoblitt, Anne G. Hummer, Meredith H. Matthews, Gerald W. Olson, and Virginia Steen-McIntyre. Thanks are also given to our Salvadoran field crew. The cooperation of four agencies of the Salvadoran government is acknowledged: Ministerio de Educación, Ministerio de Obras Públicas, Instituto Regulador de Abastecimientos, and Instituto Salvadoreño de Transformación Agrícola. This chapter benefited from readings of earlier versions by Thomas W. Killion, Payson D. Sheets, and Gene C. Wilken.

CHAPTER 10   JOSEPH W. BALL & RICHALENE G. KELSAY

Prehistoric Intrasettlement Land Use and Residual Soil Phosphate Levels in the Upper Belize Valley, Central America

IN HIS CLOSING REMARKS on the 1987 Society for American Archaeology Gardens of Prehistory symposium, William T. Sanders argued that the subsistence-related land-use patterns anciently characterizing any given prehistoric lowland Maya settlement can be inferred in most instances from simple inspectional examination of the surviving settlement remains. While agreeing in principle that the distributional patterning of a Classic period Maya settlement might provide an adequate basis on which to hypothesize the contemporary presence of a particular land-use type, we question whether this patterning alone should as yet serve as sufficient grounds on which to conclude existence of the type. We suggest instead that a conjunctive use of multiple discrete data categories still must be brought into play in addressing this issue. Artifactual data and their distribution, soil chemistry, and ancient discard patterning all ought be considered together with the surface configuration of securely dated settlement remains. The 1984–89 San Diego State University (SDSU) Buenavista Project has attempted such conjunctive studies on the sites of two morphologically and functionally distinct lowland Maya settlements of the Late Classic period (ca. A.D. 700–900). At each of these, extensive stripping excavations, distributional analyses of select artifact types and refuse deposits, and residual soil phosphate level determinations have been combined with more traditional map and test pit settlement data in an effort to reconstruct and verify the probable land-use patterns associated with them. This chapter reports on the first phase of those investigations and describes the additional work to be done in testing their results.

The 1984–89 SDSU Prehistoric Lowland Maya Community Structure (Buenavista) Project was undertaken with the aim of archaeologically documenting the synchronic sociobehavioral structure of a single, representative Late Classic lowland Maya community in terms of component groups, the activities carried out by these, and the relationships that existed among them (Ball 1983). In conjunction with these goals, project personnel have attempted to develop and test a series of hypotheses concerning the interrelationships within the community of residence location, material wealth, social status, and economic pursuits. One aspect of this effort has involved an examination of prehistoric land-use patterns around and between residential structures and other visible loci of apparently habitational nature. Our hope was to delimit individual house lots and identify other use areas associated with particular residential units and groups. To accomplish this task, we supplemented a primary research strategy of extensive stripping excavation with a program of auger and pit sampling across areas of seemingly vacant terrain. Both cultural materials and soil chemistry figured in this study, during which more than 2,000 samples have been collected and analyzed.

Two of the sites investigated as part of this program were *Buenavista del Cayo*, an architecturally defined small major center located on the uppermost east terrace of the Rio Mopan some thirteen kilometers above its confluence with the Macal to form the Belize River; and *Guerra*, a *plazuela*-focused cluster of nearly eighty isolated mounds and mound groups extending southeastward along the river from the center for a distance of approximately one kilometer (see fig. 10-1).

During 1984 and 1985, subsurface soil samples were collected systematically at both of these sites and subjected to qualitative phosphate analysis using the ring chromatography method (Eidt 1973, 1977, 1984). The analytical results suggested that substantially different patterns of open-space land use formerly characterized each site, at least insofar as these are reflected in present-day soil phosphate levels. We then car-

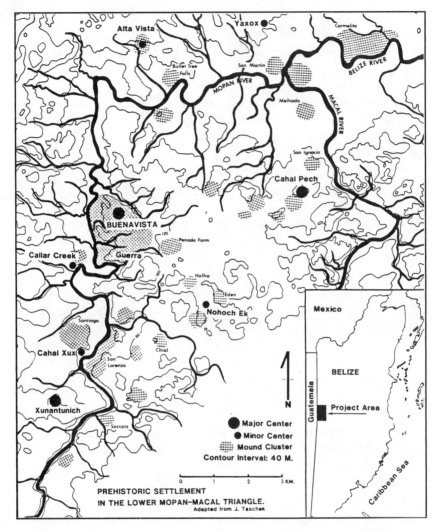

Figure 10-1. Buenavista de Cayo-Guerra Study Area

ried out two series of follow-up excavations for the purpose of examining the subsurface associations of the phosphate samples and clarifying their potential cultural significance. Full evaluation of these excavations remains to be completed, but those data that are now available are incorporated as appropriate below.

PROJECT METHODOLOGY

The procedures employed for sample collection and analysis were based on those introduced and developed by Robert C. Eidt (1973, 1977) with modifications added by Kelsay (1983). These procedures involved the use of both mound-focused radial sampling transects (see fig. 10-2) and intermound sampling traverses. Soil samples were obtained at one- or two-meter intervals along these from a constant depth of thirty centimeters.[1] In a few instances, it was not possible to obtain samples due to the existence of subsurface obstructions. The locations of all such "missing data" are indicated on the accompanying graphics.

Sample processing was carried out using the ring chromatography method (Eidt 1977, 1984).[2] Because this test yields results in terms of relative soil phosphate values, its effective use for archaeological purposes requires first establishing a local natural off-site soil phosphate standard. On-site test results are then measured against this standard to determine localized enhancement, depletion, or neutrality. For Buenavista and Guerra, local natural standards were established based on five thirty-centimeter depth samples drawn from locales manifesting no surface evidence of ancient settlement situated approximately two kilometers northeast, two kilometers east, and two kilometers southeast of Guerra's outer edge. In actual point of fact, the local soils are generally deficient in naturally occurring fixed phosphates. Our off-site tests yielded readings of *very low* (about 25 pounds or less of phosphorous per seven-inch deep acre) to *low* (about 50 pounds per seven-inch acre). On-site results ranged from *medium* (about 75 pounds per seven-inch acre) to *very high* (200 + pounds per acre).[3]

The off-site low to very low values were used to establish our reference standards, and samples collected from on-site were measured against these using a Hellige-Truog fixed standard colorimetric phos-

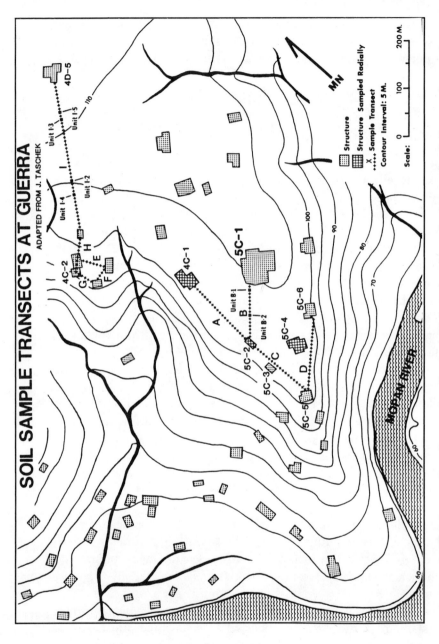

Figure 10-2. Guerra Site. Conventionalized settlement plan shows locations of radially tested mound groups, intergroup phosphate sampling transects, and follow-up excavations.

PREHISTORIC INTRASETTLEMENT LAND USE

phorous comparator and a *Munsell Book of Color.*[4] Accurate interpretation of test results has been complicated by recent land-use patterns in the site vicinity.

Originally forested, both Guerra and Buenavista were cleared in the late 1960s for use as pasture to graze cattle, sheep, and horses. Livestock will add measurably to soil phosphate content but to a lesser extent than do human activities. Conversely, the harvesting of plants will bring about marked deficiencies in soil phosphate levels (Eidt 1984:28–31). In view of these potential complications, we would emphasize that the off-site test loci were selected with attention to comparability in recent land use to both Guerra and Buenavista and that these two themselves have had a common use-history since at least the turn of the century.

During both 1984 and 1985, follow-up test excavations were carried out along several of the soil sampling transects at locations corresponding to different values on the relative phosphate level scale. These units were excavated to sterile soil where artifactual material was present or to an equivalent depth where not. The resulting data were then combined with those from the phosphate tests to aid in their interpretation.

In lateral extent, individual tests were begun as one-by-two- or two-by-two-meter units and expanded in a few select cases into three-by-four-meter clearing operations. Our intent was to expose the minimum horizontal areas necessary to make possible both chronological and functional evaluations of the depositional contexts associated with specific phosphate levels. The few lateral expansions carried out involved efforts to clarify or verify the findings from more restricted test units. Still being evaluated are data from four three-meter-wide strip excavations immediately flanking and paralleling phosphate sampling transects I, C, and D at Guerra and A at Buenavista (see figs. 10-2 and 10-6). These data will be presented and discussed on a future occasion.

THE 1984–1985 SAMPLING PROGRAM: PART 1—THE MODEL

Guerra

In the terminology of the 1977 School of American Research advanced seminar on lowland Maya settlement patterns (Ashmore

1981a), the Guerra site comprises an extensive *plazuela*-focused cluster of individual mounds and mound groups (fig. 10-2). Spreading out over some 160 hectares of gently sloping riverside terracelands, Guerra is contiguous with the Buenavista center on its north and for this reason has been characterized by us as "suburban." Extensive stripping excavations and supplementary trench sampling have made possible its identification as an agriculturally oriented settlement site (Taschek and Ball 1986). Although the scene of multiple discrete occupations, Guerra's visible surface remains date exclusively to the Late Classic period (ca. A.D. 700–900) and have yielded ceramic and obsidian hydration-based evidence of residential use during the closing decades of that span.

Physically, Guerra consists of seventy-eight isolated mounds and mound groups together with one larger complex of six mounds atop an elevated, masonry-faced platform (Group 5C-1). Both the solitary units and the groupings of two, three, or four mounds sit atop low cobblestone platforms. Originally leveled and surfaced with layers of tamped earth and clay, these platforms extend outward from beneath the mounds to provide structurally defined areas of adjacent or enclosed open space. Again following the 1977 SAR seminar (Ashmore 1981b), we have labeled these elevated clear areas *patios* and identify the composite platform-and-mound units as *patio groups.* Extensive stripping and complementary trench tests have allowed identification of the mounds as onetime habitations and/or kitchens and/or domestic storage buildings (Taschek and Ball 1986). They likewise have shown the patio areas to have been the scenes of tool production/maintenance/ repair, food processing, and other household-related maintenance activities.[5] Potsherd densities across three of these patios averaged about 5 kilograms per cubic meter, with a range of from roughly 2.6 to 7.2 kilograms per cubic meter.[6] Other artifacts and debris recovered are consistent with patterns of occasional loss and low-density discard remaindering.

Just off the edges of and bordering these patios are concentrated bands of "hard" (inorganic) trash within which sherd densities increase to an average of 11.8 kilograms per cubic meter (range = 5.3 to 18.9). Beyond these bands, sherd and other artifact densities drop to an average of less than 600 grams per cubic meter except at loci of buried earlier (pre–Late Classic) features.

A map of Guerra reveals a dispersed settlement of what we have interpreted to be some eighty discrete household units distributed over approximately 160 hectares of land (see fig. 10-2). Each hypothesized residence complex (or mound group) is associated with between roughly 0.8 and 2.8 hectares of surrounding open space.[7] Superficially, this space is today indeed "open" and comprises a classic example of what settlement archaeologists have come to refer to as "vacant terrain." Whether or not it was so anciently, and in either case to what use it might have been put, are questions our research has attempted to address. Surface features are entirely absent, and off-mound artifactual materials are extremely rare. In almost all cases, these latter clearly owe their presence to natural erosional or depositional forces.

Employing Sanders's suggested inspectional approach, an intrasettlement land-use pattern involving the onetime presence of both house-lot gardens and infield farm plots might quite legitimately be posited for Guerra. Indeed, drawing on personal observations made during 1975 of present-day land-use patterns within the highland Chiapas Tzeltal Maya community of Amatenango del Valle, the authors had themselves made just such an assumption. Our hypothesized model also envisioned circumsettlement infield plots and outfields extending eastward away from the settlement and river.

To date, neither aerial nor ground surveys have produced any data contravening such a model. Rather than adopt it on the prima facie evidence of settlement layout alone, however, we elected to test it using other than traditional surface survey methods. In considering the possible means through which we might accomplish this test, we first sought an appropriate general ethnographic model from which to educe the probable material effects or correlates of analogous present-day land-use patterns.

A detailed land-use model of a closely analogous ethnographic system has recently been described for Tamshiyacu, a rural peasant agricultural community on the upper Amazon River in northeast Peru (Hiraoka 1986). Stripped of its living populace, this riverside Amazonian village presents a nearly perfect modern counterpart to Guerra in settlement layout, microenvironmental situation, broader ecological setting, and general subsistence technology (Hiraoka 1986:355–57, figs. 2 and 3). Its subsistence-oriented land use is diversified and incorporates a mixture of house-lot gardens, infield plots, and outfields, each involving a range of both agricultural and horticultural cultivation. Not every household engages in all three strategies to the same degree, nor even in all three. This fact accords well with Killion's findings from southern Veracruz (1987a, 1987b, this volume). There also considerable intrasettlement variation involving subsistence strategies and associated land-use patterns is to be found within a suitably analogous subsistence-settlement system. The Guerra map suggests the likely existence of comparable variability within that ancient community.

Recent ethnoarchaeological studies among the Tzeltal and Chuj Maya of the Guatemala-Chiapas highlands (Hayden and Cannon 1983, 1984) and among Tuxteco peasant agriculturalists in southern Veracruz (Killion 1987a, this volume) have provided pertinent complementary data concerning the treatment and disposition of household refuse within ethnographic communities of this general type within individual house lots and across the interstitial settlement space between them. Practices including selective trash removal, separation, disposal segregation, the accumulation of concentrated refuse deposits, and the differential fertilization (chemical alteration) of intrasettlement soils with organic household wastes all have been reported or implied. A remarkably high degree of regularity occurs in the spatial patterning and physical character of the "archaeological" records resulting from these activities (compare, for example, Hayden and Cannon 1983, especially fig. 5, and Killion, this volume, especially figs. 6-6 and 6-7), and we believe the ethnographic cases suggest at least one viable alternative through which to approach the identification of analogous prehistoric land-use patterns.[8]

The generalized model of intrasettlement land use and refuse treatment that emerges from the ethnographic cases comprises a series of contiguous concentric discard and household activity zones centered on discrete residential structural cores. In each of these, a *structural core* or *residence complex* consisting of one or more buildings for sleeping, storage, and/or kitchen activities is adjoined or surrounded by a well-kept open space where neither "hard" (inorganic) trash nor organic refuse is allowed to accumulate *(zone a)*. Both the structural core and the open space are scenes of routine household manufacturing, maintenance, and food-processing activities from which trash regularly is swept or removed. The dimensions and surface character of this area may vary considerably, but the area can be defined heuristically by the accumulation along its perimeter of heavy inorganic trash concentrations (Hayden and Cannon 1983; Killion, this volume).

Beyond the circumresidence clear zone and its edging band of concentrated refuse *(zone b)* extend the house-lot gardens *(zone c)* and infield plots *(zone d)*. Hayden's data suggest that the bulk of organic wastes is most likely to provide enrichment for the house-lot garden (Hayden and Cannon 1983:126–30). This pattern corresponds well with Hiraoka's (1986:366) findings at Tamshiyacu. In both cases, some limited organic enrichment of intrasettlement infields also occurs, but in neither does it constitute a regular or generally expectable practice.

Our archaeological data clearly and unequivocally correspond with the ethnographic findings of Killion (1987a, 1990, this volume) and Hayden (Hayden and Cannon 1983, esp. 126–40, fig. 5) concerning refuse treatment. These findings might quite reasonably serve as a basis for interpretive statements regarding trash disposal and related spatial management in the prehistoric community. We believe that equally sound inferential conclusions regarding subsistence-related land use can be drawn from patterned variations in on-site (intrasettlement) soil phosphate levels where these exist.

Operationally, the generalized ethnographic model described is most usefully restated as a series of expectations or predictions regarding the distribution of both hard trash and soil phosphate concentrations across a given site. Our expectations concerning these at Guerra were as follows.

Immediately adjoining or enclosed by each archaeological "structural core," we anticipated an initial clear zone of varying extent free of accumulated refuse and low in residual phosphate content *(zone a)*.[9] Bounding this clear zone at some distance out, we expected a heavily concentrated band of hard trash possibly also incorporating some limited organic debris *(zone b)*. Beyond this zone, our expectations were for a zone of soils with little to no contemporary artifactual content but with consistently if somewhat variably elevated ("enriched") phos-

243

phate levels *(zone c)*. A possible fourth interstitial zone *(d)* was allowed for within which phosphate levels might decline to an intrasettlement low from which they would eventually rise again in a mirror-image pattern toward another structural core.

Occasional phosphate and trash "hotspots" defining favored urinary locations and secondary dumping areas were expected in zone c. Ideally, we also hoped to find evidence of Classic period house-lot boundary walls like those recently identified farther north at Coba (Benavides and Manzanilla 1985) and around Becan (Thomas 1981). Thus far, however, we have been unsuccessful in establishing the existence of any such similar features at Guerra. We suspect this absence of evidence more probably reflects the use of perishable fencing materials at Guerra than it does any truly fundamental differences in the land-tenure patterns characterizing the three regions.

To provide something of an archaeological check on our findings at Guerra, we also collected a second series of samples for residual phosphate evaluation from a contemporaneously occupied but morphologically quite dissimilar site, Buenavista del Cayo.

Buenavista

A little over one kilometer north of its southern edge, the Guerra settlement adjoins an extensive concentration of monumental platform architecture, range structures, and architecturally defined plazas. Buenavista del Cayo is a small regional center of the level 8 or 9 category in Norman Hammond's (1975) 10-level typology of Classic period lowland Maya centers. At the heart of a territorial realm estimated to have encompassed some roughly 175 square kilometers during its Classic period florescence, Buenavista counts among its remains two ballcourts, an acropolis-palace, and at least ten additional courtyard groups. At least one of its pyramidal platform structures stands today still to a height of more than twenty-one meters. Altogether some twelve hectares of contiguous monumental architecture are spread over approximately eighteen hectares of space.

Excavations carried out at Buenavista between 1984 and 1989 demonstrate it to have been a narrowly focused multifunctional or "urban" center of the "regal-ritual" type (Fox 1977; Taschek and Ball 1987; Ball and Taschek 1989). Our data document its role as a zonal node of public ceremonies, high-status residence, high-status funerary rites, and high-status residence-sited adjudicative-administrative activities.[10]

The likely presence of attached specialist craftsmen and artists is attested to by the discovery of palace-associated workshops engaged in the production of polychrome-painted pottery and engraved bone ornaments (Ball 1991). At the same time, beyond the possible siting of periodic festivals and whatever secondary economic activities might have accompanied these, little evidence exists to support any role as a focus for distributive activities or events any more extensive, intensive, or regular in occurrence than those taking place in surrounding rural settlements.

In observed morphology and inferred function, Buenavista plainly differs radically from Guerra. Theoretically the two should differ as well as in intrasite land-use patterns and the corresponding distributions of residual soil phosphates. Our sampling program at Buenavista had as its goals the testing of this proposition and a tentative evaluation of interstructural land-use patterns within the inspectionally urban site. As detailed below, our initial findings suggest a significant departure between the two sites in ways consonant with the differing functional roles suggested for them on purely inspectional grounds. In the following section, our field testing procedures and analytical findings are described. Interpretation of these in land-use terms is reserved for the subsequent final section.

THE 1984–1985 SAMPLING PROGRAM: PART 2—PROCEDURES AND RESULTS

We selected three structural mound groups at Guerra for radial sampling in an effort to explore individual house-lot sizes and potential associated use areas. Samples were collected at one-meter intervals along multiple transects radiating outward from the visible edges of each mound for distances of thirteen meters (see fig. 10-3). In addition, nine vacant terrain transects were tested to identify possible "invisible

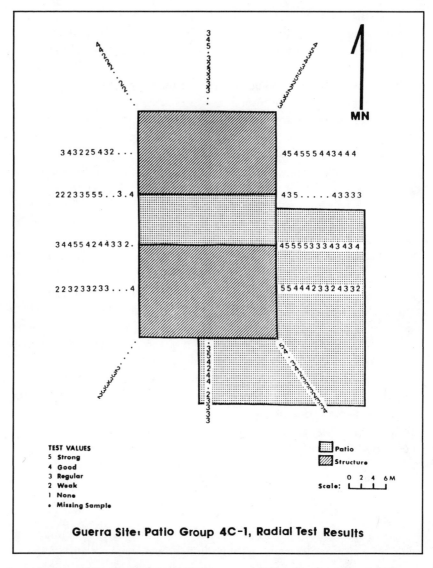

Figure 10-3. Patio Group 4C-1, Guerra. Schematic plan shows phosphate radial test results.

houses" or other activity areas not evidenced by any readily apparent surface indications. The empty terrain traverses were sampled at two-meter intervals.

Following processing of these samples, some one-by-two- and two-by-two-meter follow-up tests were excavated along several of the transects at points corresponding to oppository value extremes on our relative test scale. Each unit was excavated to sterile soil where artifactual material was present or to an equivalent depth where not. The recovered materials were analyzed and the data combined with those from the phosphate tests to aid in their interpretation.

Radial Tests

Patio group 5C-2 consists of a single small building and its frontally adjoining patio (see fig. 10-2). Test results indicate two extensive areas of heavy phosphate concentration with several isolated samples reading "4." Of the two larger areas of concentration, one lies seven meters to the west and downslope from the structure. The second, approximately seven meters in extent, begins four meters to its northeast in a superficially "empty" area of interstructural open space. The isolates lie between from four to seven meters from the structure in varying directions. With only one exception, the area immediately adjacent to the structure exhibits only a slight elevation in phosphate level, if any. The single exception is within one meter of the rear wall of the structure.

At 5C-4, a four-structure patio group, a downslope area of heavy phosphate concentration occurs again, this time to the northwest. A second zone of concentration begins in an area of level open terrain approximately six meters to the southeast of the complex. As at 5C-2, several isolated areas of phosphate-enriched soil occur between roughly four and seven meters of the structural core in varying directions. Areas immediately adjacent to the patio group once again appear to be enriched only very slightly if at all.

The third group sampled, 4C-1, comprises two rectangular buildings facing each other across a cobblestone platform-patio (fig. 10-3). Although several points were not sampled due to the location of major excavation backdirt piles, three areas exhibiting highly elevated levels of fixed phosphate are nonetheless clearly discernible. One lies to the west and north, downslope of the complex toward a ravine. The second, extending some seven meters, begins in an open area approximately five meters northeast of the group. The third and most extensive area of heavy concentration lies along the east and southeast side of the patio and southern structure. Upon investigation, it was found that a debris-laden layer of cobbles, like those composing the patio, underlay this area at a depth of thirty to forty centimeters. This layer ultimately was identified as the compressed and smeared remains of a fifth- or sixth-century patio group underlying the eighth-century surface complex.

No one of the foregoing patterns alone provides an especially clear or convincing reflection of ancient land-use practices. A faint but definite ghost image does become discernible when they are considered in concert, however.

Least ambiguous are the several areas of phosphate-enriched soil located a short distance away and downslope from each patio group. Hayden and Cannon (1983) have discussed at some length the quite conscious and deliberate use of downslope areas, ravines, and streams within a two-minute walk of the residence complex as final disposal loci both for organic household sweepings and the smaller inorganic trash among the present-day highland Maya. We feel little doubt that the several downslope "hotspots" detected at Guerra are of similar origin.

A second recurrent series of phosphate-elevated zones and spots extending outward from between roughly four to twelve meters from the edge of each structural core is better appraised in concert with the transect results described below. Together these data clearly demarcate a belt of enrichment surrounding each structural core from a point just beyond the outer edge of its zone b hard-refuse band to a distance of roughly ten to twelve meters out. The results are remarkably consistent from group to group, and we believe they define the limits of house-lot gardens organically enriched with household wastes. The enrichment pattern is consistent with one of periodic small-scale dumping such as would be expected in this case (cf. Hayden and Cannon 1983). The house lots so defined would range in average area from roughly 0.08 to 0.1 hectare.[11]

Most ambiguous of the radial test data are the occasional phosphate hotspots recorded in immediate proximity to the exterior rear or side walls of several buildings. These spots could represent "bleeding" of kitchen activity organics from within these structures or building dripline sludge deposits. Alternatively, such hotspots may reflect ancient human and/or canine urinary activity.

Vacant Terrain Tests

In addition to the foregoing investigations, nine strips of structurally featureless open space also were sampled for phosphate-level determination (fig. 10-2). As noted above, these strips involved areas either totally devoid of artifactual surface materials or, in the case of one dispersed mound cluster (group 4C-2), lightly to moderately sprinkled with sherds and other debris. Of nine transects, five ran between widely separated structural groups while four ran between the units of a single, seemingly loosely integrated albeit still somewhat dispersed mound cluster.[12]

Vacant terrain phosphate transect A was run between patio groups 4C-1 and 5C-2. Eighty-seven percent of the samples indicated some degree of phosphate enrichment of the soil had taken place, although approximately 58 percent reflected only moderately elevated levels with a value of "3." The remaining 29 percent registered "4" *(enhanced)* on the test scale. These were clustered into two groups at opposite ends of the transect within the ten to twelve meters immediately adjoining the patio group structural cores.

A two-by-two-meter test unit (A-1) was excavated at a point located twenty-six meters north of 5C-2 in the midst of an approximately eight-meters-broad area registering "3" on the scale. Although no sherd or lithic debris was present on the surface at this point, subsurface materials appeared and became increasingly richer and more abundant from about fifteen to twenty centimeters below surface to bedrock at minus forty-five centimeters. The cultural materials recovered from this unit included numerous medium to large, moderately eroded late Early and Middle Classic period potsherds; an assortment of chipped stone tools, flakes, fragments, and shatter, much of it exhibiting thermal alteration and use-wear; numerous chunks of baked daub; and a light peppering of carbonized wood, river snail fragments, and obsidian blade segments. The materials appeared to represent Middle Classic (ca. A.D. 550–650) household refuse, and ultimate expansion of unit A-1 into a three-by-four-meter exposure revealed them to be the remains of a small residential building of that date.

Transect B ran between patio group 5C-2 and *plazuela* group 5C-1 (figs. 10-2 and 10-4). Of the thirty-three samples collected, only one indicated no increase in phosphate above naturally occurring levels. Thirty-nine percent of the samples produced moderate values of "3," while 30 percent yielded values of "4" and 27 percent registered "5." As can be seen in figure 10-4, the areas producing samples with significantly elevated readings ("4" or "5") form three loosely defined zones. Two of these occur immediately adjacent to and extend outward from the two structural cores for varying distances of roughly ten and twenty-six meters. Quite unremarkably, the greater distance is associated with the considerably larger 5C-1 *plazuela* group. The third is located in the interstitial space between the two groups approximately twenty to thirty meters out from the smaller and just short of halfway between them.

Two test units were excavated along transect B. The first, B-1, was placed where a strong reading of "5" had been obtained some sixteen meters out from the edge of *plazuela* 5C-1. No artifactual material was present on the surface at this point. The unit, a two-by-two-meter test supplemented in 1988 by a three-by-three-meter excavation, yielded a high-density mélange of small, badly broken-up pottery sherds, obsidian blade segments and reduction shatter, broken chert flake-tools, and extensive chert secondary reduction debris. We believe this to be secondary refuse (Schiffer 1976, 1977), most probably redeposited from the nearby *plazuela* group. An unusually high frequency of imported obsidian and the occurrence of other exotic materials (marine shell scrap, a hematite mosaic inlay plaque) in the deposit support its derivation from an elevated-status social context. We suggest that it probably represents a discrete, final discard dump as described by Hayden and Cannon (1983: esp. 139–40, 148–49, fig. 5) but whether within or outside the house-lot bounds we remain uncertain.

In an effort to obtain comparative data for a point low in residual phosphates, a second unit, B-2, was situated to follow up on the single

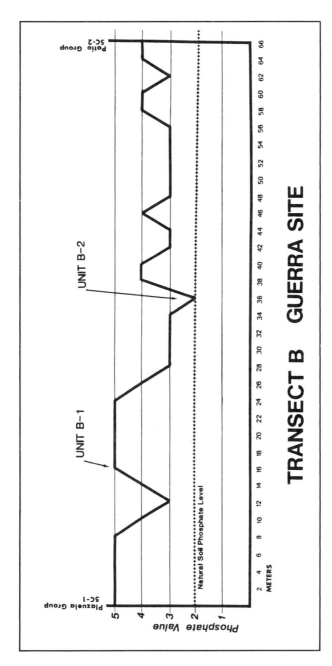

Figure 10-4. Soil Phosphate Levels and Follow-up Test Locations; Soil Phosphate Sampling Transect B, Guerra Site

transect B sample indicating an unelevated phosphate level (see figs. 10-2 and 10-4). Located approximately midway between the two mound groups, this unit encountered a low-density deposit compositionally not unlike B-1 in ceramic and chipped lithic content. A few probable Late Classic modes and some possible Middle Formative ones as well were present among the small, badly eroded and rounded sherds. We suspect this deposit to represent yet another discrete dumping locus, but the low level of residual phosphate would seem to suggest that the trash dumped here lacked any appreciable amounts of organic waste. Possibly this was a hard-refuse accumulation at or along the boundary dividing the two Classic period house lots. Alternatively it could represent an extra-house-lot secondary dumping site such as are found commonly today in highland Maya villages (Hayden and Cannon 1983). Again, the materials recovered might be the remains of either a severely "smeared" residential unit or a secondary refuse deposit substantially earlier in date than the Late Classic Guerra settlement. Thus far, the interpretive significance of the test B-2 data remains elusive.

Transect I ran between patio groups 4C-2 and 4D-5 (fig. 10-2). Seventy percent of the samples indicated elevated phosphate levels. Of these, 66 percent yielded values of "3," 15 percent values of "4," and the remainder values of "5" (fig. 10-5). There was, by this point, a predictable concentration of highly elevated levels within the first ten meters adjacent to group 4D-5, but otherwise the distribution of "highs" and "lows" did not seem to follow any readily understandable pattern. Four test units were excavated along the transect, two at places exhibiting no enrichment of the soil and two at locations registering high residual phosphate levels.

Test unit I-2 yielded only a few, tiny, severely eroded pottery crumbs and some small chert flakes and shatter. This small yield and the absence of any chemical evidence for phosphate enrichment would seem to indicate essentially unaltered open space. The location of unit I-2 (see fig. 10-2) makes it probable that this was interstitial intrasettlement space (zone d), but whether representing unused space or cultivated plots (infields) is not as yet readily apparent. We plan future excavations in this and other comparable areas to resolve this question.

Test unit I-5 encountered a situation superficially much like that described for test B-2. Here again considerable artifactual debris was

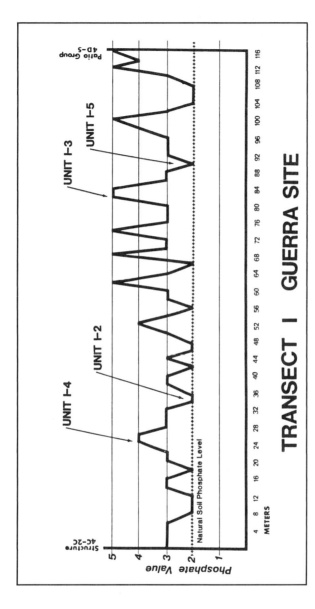

Figure 10-5. Soil Phosphate Levels and Follow-up Test Locations; Soil Phosphate Sampling Transect I, Guerra Site

present without evidence of organic refuse. Although the sherds present were badly eroded, most appeared assignable to the late Middle Formative period (ca. 600–400 B.C.). Further excavation revealed the deposit to be the compressed and smeared remains of a late Middle Formative house.

Units I-3 and I-4 encountered depositional situations essentially the reverse of those found in B-2 and I-5. Both units produced nearly equal numbers of badly eroded small sherds and lithic debris. Frequencies were considerably lower than those from unit B-2 or I-5, yet the I-3 and I-4 samples produced strong phosphate readings of "5" on the test scale. The presence of organic wastes in substantial amounts is thus indicated for these loci. What is less than clear is their significance and origin. Tentatively, we suggest they reflect the Classic period existence of intrasettlement infields (our zone d) fertilized with household organic refuse. We will investigate this proposition through more extensive stripping excavations in the future.

Transects E, F, G, and H were run between the structures of mound cluster 4C-2, a putative extended-family residential complex (see fig. 10-2). Phosphate levels recorded along these transects were uniformly low to moderately elevated other than within the first two meters out from each building and across a slightly mounded area at the northwest corner of the cluster. Two test units were placed along transect G. Unit G-1 was situated at the northwest corner intersection of transects G and H where a very high phosphate level reading of "5" had been obtained. The artifactual materials recovered compared closely to those from transect test A-1 and ultimately were identified as a house ruin of somewhat earlier date than the more pronounced mounds making up the cluster.[13]

Follow-up test G-2 was located at a point showing no evidence of residual phosphates on the west edge of the group 4C-2 structural core. The recovered materials included chipped stone tools, fragments, production/maintenance debris, and numerous very small pottery sherds. In concert these appear likely to represent redeposited household refuse unaccompanied by any significant quantities of organic waste. This is probably the zone b band of hard refuse bordering and defining the limits of the Late Classic residence complex.

Buenavista

Of the four transects run within the Buenavista center, three (transects A, B, and C) sampled apparent vacant terrain between visible structural groups in order to acquire data, making possible a comparison of land-use practices between the organizational center and its associated residential suburbs (see fig. 10-6).[14] Transect D sampled an area lying between structure 6D-1, a single-family residence unit, and a nearby feature tentatively identified as the dump site for a high-volume woodworking shop (Kelsay 1985). These last samples remain to be analyzed.

Transects A, B, and C uniformly yielded soil phosphate patterns markedly different from those found at Guerra (fig. 10-7). Eighty percent of our samples indicated either unaltered or actually depleted levels, results in agreement with an anciently unenriched, recently harvested use history. In essence, no evidence exists for the occurrence of soil-altering human activities in the area between the four urban residential groups. Two- to three-meter bands of hard refuse edge each group. Beyond this point, the buried land surface appears clear of artifactual debris and free of organic enrichment. No evidence of intervening plaster pavement has been found, and we are left with a hypothetical landscape of dispersed residence groups separated by expanses of unfertilized and so presumably unfarmed open space ringing the Buenavista center. Within the central twelve hectares of the center, individual residences, residence complexes, and public buildings were separated by extensive open expanses of plaster pavement. Obviously no significant nonornamental agricultural activities took place within this zone. Trash disposal, on the other hand, seems to have followed much the same pattern as elsewhere. Refuse piled up just off the platform edges of even the highest-level elite residence groups, and the only significant difference in its management would appear to have been its occasional large-scale use as structural fill in the center's extensive construction activities. The overall picture obtained from within the Buenavista center is one that contrasts sharply with the use patterns revealed at Guerra.

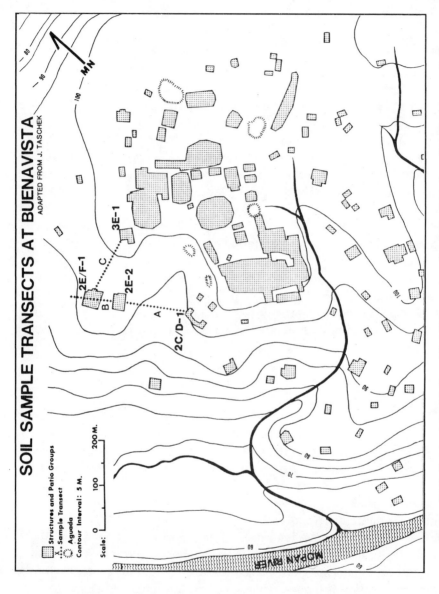

Figure 10-6. Buenavista del Cayo Site; Conventionalized Site Plan Locating Intergroup Soil Phosphate Sampling Transects

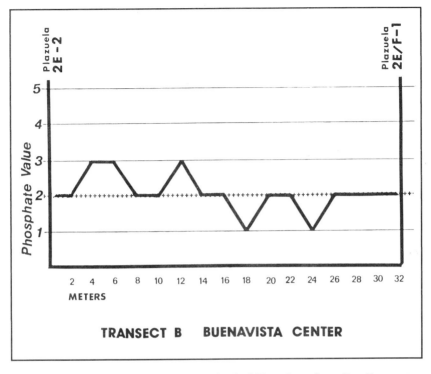

Figure 10-7. Soil Phosphate Levels; Soil Phosphate Sampling Transect B, Buenavista Site

CONCLUSIONS

A systematic examination of residual soil phosphate levels at Guerra reveals a distribution congruent with the probable onetime existence of an intrasettlement land-use pattern involving house-lot gardens and/or interstitial infields. Our identification of this pattern rests on multiple analogies with a complex ethnoarchaeological model of agricultural land use, refuse disposal, and spatial management within a series of ecologically and technologically comparable modern farming communities. Clearly, as William Sanders correctly observed, simple inspection of the Guerra settlement map is likely to lead to similar conclusions. We believe, however, that the field and laboratory studies here reported serve to help confirm the existence of such a pattern rather than simply to suggest its likely occurrence. On their basis, future inferences regarding other comparable sites can be made with somewhat greater confidence.

As do their rural village descendants today (Hayden and Cannon 1983:126–40), the prehistoric Maya appear to have separated their disposable wastes into a number of different inorganic and organic categories. Again as today, these were segregated in deposition, the one body going into trash piles, the other serving to fertilize agricultural plots. At Guerra, the archaeological distributions of hard refuse and phosphate-enriched anthrosols suggest a Classic period landscape of interresidential fields fertilized discretely but regularly with organic household wastes. What remain uncertain, however, are the sizes and boundaries of individual house lots. Thus we are unable to state with real certainty whether these intrasettlement plots were residential ("kitchen") gardens or infields or both. To a large extent, this problem is due to the absence of easily identified stone-wall house-lot boundaries within the Guerra settlement. We also remain unable to specify what kinds of crops were grown at Guerra. In an effort to resolve these questions, we have initiated work along two additional avenues of investigation.

First, during 1988 we carried out a new series of vacant terrain stripping trench excavations at Guerra. These consisted of three three-meter-wide trenches excavated down to the Late Classic period land surface and run from mound group to mound group across the open terrain paralleling phosphate transects I, C, and D. Two additional new transects accompanied by flanking trenches also were excavated at a second suburban settlement site located about one kilometer northeast of the Buenavista center. Our hope is that more refined data from these investigations will enable us to identify currently "invisible" house-lot boundaries. We are especially interested in the distribution of discontinuous concentrations of hard trash and their spatial relationships to structural features and areas of more diffuse, continuous refuse accumulation. We believe on analogical grounds that these data may help resolve our problem (see Killion, this volume; Hayden and Cannon 1983; endnote 8, this chapter).

As to the question of crop type(s), a select series of soil samples from

Guerra currently is undergoing qualitative fractionation analysis at the CALTRANS Soils Laboratory, San Diego. Recent studies by Eidt (1984) have demonstrated the potential of fractionation analysis to identify the cultigens responsible for anthrosol phosphates.

Although more tenuously based as yet, a strikingly different pattern appears to have characterized interresidence space within the urban center, Buenavista. There, no evidence has been found to support the existence of ancient intrasettlement fields or gardens although trash disposal still appears to have occurred in immediate proximity to each residence unit. Future work will include expanding our data on intrasettlement land-use patterns within the Buenavista center through follow-up excavations and transect-flanking trenches like those used at Guerra.

Returning, finally, once more to the empirical testing of hypotheses concerning prehistoric land use, the ring chromatography method of residual soil phosphate analysis can serve as one effective aid in the detection and spatial definition of subsurface cultural remains or localized areas of ancient activity involving the deposition of organic wastes. In our experience, however, it has not been able to differentiate among or functionally characterize these in the absence of follow-up excavations. It is not a foolproof means for identifying either middens or "invisible" houses as some have suggested. In all cases, the determinant factor in its effective operability is the concentrated presence of anciently deposited organic matter. Used thoughtfully, the method can provide the archaeologist working in the Central American tropics as elsewhere with a powerful investigative approach to questions of prehistoric extraresidential land use. The demonstration of both its real utility and its equally important limitations quite possibly has been the greatest value of the method's application at Guerra.

ACKNOWLEDGMENTS

This chapter represents a substantially altered version of a paper presented at the 1987 annual meeting of the Society for American Archaeology. The research reported was supported in part by a grant from the National Science Foundation (no. BNS-8310677). Permission to carry out fieldwork in Belize was graciously extended by Harriot W. Topsey, Commissioner of the Department of Archaeology, and by Antonio Guerra on whose lands the Buenavista and Guerra sites are located. Project field director Jennifer T. Taschek greatly facilitated both the soil sampling and the follow-up excavation trenching through her willing diversion of workers and equipment from more "traditional" archaeological activities. Finally, Thomas W. Killion's extensive comments and suggestions greatly influenced the improvement that we hope this chapter represents over our earlier SAA paper. To each of these individuals we wish to express our heartfelt appreciation. At the same time, we take full responsibility for any errors of fact or interpretation that may flaw this study.

NOTES

1. Samples were drawn using a manually operated four-inch-diameter molybdenum steel soil auger. Each sample was sealed in a sterile plastic bag and labeled for storage and later processing. Processing took place in the project field laboratory during 1984 but in 1985 and after was deferred until the samples could be returned to San Diego. This change resulted from our inability to control daily temperature levels inside our Belize field lab. These commonly were above 29° C and frequently exceeded 32°. Temperatures above 27° C cause the ammonium molybdate/hydrochloric acid solution (see n. 2) to become unstable and can adversely affect test results.

2. Ring chromatography processing involves the use of two reagents. The first, Reagent A, is an extracting solution and consists of 35 milliliters of 5N hydrochloric acid mixed with 5 grams of ammonium molybdate dissolved in 100 milliliters of cold distilled water. The second, Reagent B, is a reducing solution consisting of 0.5 gram ascorbic acid dissolved in 100 milliliters of distilled water. A small amount of soil—approximately 5 grams—is placed on a seven-centimeter-diameter ashless (phosphate-free) filter paper. Two drops of Reagent A are added to the sample. Thirty seconds later, two drops of Reagent B are added. If phosphates are present in the sample, a blue ring will form around the sample. Readings are taken at a uniform two minutes after the addition of Reagent B. A value is assigned to the relative level of phosphate indicated based on: (a) the time of the blue ring's appearance, (b) the percentage of the ring present, and (c) the radial width or depth of the ring. Standardly

used scale values are: 1—none, 2—weak, 3—average, 4—enhanced, 5—strong. For a thorough discussion and history of the ring chromatography method, the reader is referred to Eidt 1984: esp. 33–44.

3. The native soils in the area of the Buenavista and Guerra sites are limestone-based clays of the Yaxa and Melinda suites with a pH of about 6.0 and only 4 to 5 percent organic material in the topsoil. Drainage is slow and soil structure poor (Wright et al. 1959:88).

4. All basic determinations were made using a No. 697 Hellige-Truog Combination Soil Tester set and its fixed standard colorimetric comparators. These were supplemented as necessary with appropriate pages from the *Munsell Book of Color*. Individual test color results were recorded for reference in Hellige-Truog or Munsell form as each discrete sample was analyzed.

5. All such ascriptions are based on extensive in situ data including broken and discarded or lost artifactual materials and primary refuse (see Schiffer 1976, 1977) rather than on questionable "associated fill" materials as is generally so in the case of prehistoric lowland Maya settlement studies (for important exceptions to this see Haviland 1981, 1985). Evidence documenting both community-wide domestic and economic activities and more confined craft production was recovered. These data will be presented in detail in several future reports.

6. These figures represent standardized extrapolations from cultural deposits and archaeological sampling units of widely varying depths and extents. Although sherd densities traditionally have been presented as numerical frequencies, we have found these unreliable and invalid as a basis for comparative statements. Numerous factors (among them paste, temper, and firing attributes; chronology; original depositional circumstances; postdepositional events; use-life functions; depositional causes; climatic exposures; and even archaeological recovery techniques) can affect numerical frequencies to the extent that the materials from two discrete deposits rarely are truly comparable. To address this somewhat, we have adopted volumetric mass as our standard of measurement and comparison. While the aforementioned factors and others may severely affect numerical frequencies, volumetric weights and density measures based on them will remain relatively constant under all but the most extreme circumstances.

7. Walled fields of probable Late Postclassic date recorded on Cozumel Island range in size from under 0.1 to just over 1.0 hectare (Freidel 1976).

8. One interesting and archaeologically significant disparity does exist between Hayden's highland Chiapas-Guatemala findings and those of Killion in southern Veracruz. This disparity concerns the disposal of refuse in house-lot gardens. Killion (this volume) found house-lot gardens to be repositories of diffuse or broadcast trash. Hayden and Cannon (1983), in contrast, report the

house-lot garden occurrence of inorganic trash only in discrete dumps if at all. More importantly, Killion has identified a strong and direct relationship between house-lot size and the spatial management of trash, whereas Hayden and Cannon report that not size but the kinds of activities taking place within specific house lots seem to be the major determinant of refuse patterns. We suspect technomic rather than "regional" difference factors to be at work here, and we believe the Chiapas-Guatemala data are more likely to apply to our situation for this reason. Clearly this is an area warranting further ethnoarchaeological examination.

9. The possible presence of use-related casual trash items and occasional urinary phosphate "hotspots" was allowed for.

10. The Buenavista project is employing a provisional ranked social typology based on residence situation, residence character, demonstrated access to exotic materials (such as marine shell or jadeite), and demonstrated access to high energy investment local goods (such as polychromed pottery or carved bone jewelry). Currently recognized strata and hypothesized relationships are: upper-level urban elite, upper-level rural elite, lower-level urban elite, lower-level rural elite, urban middle, suburban-rural middle, urban commoners, suburban-rural commoners, and peasantry.

11. Actual house-lot sizes remain undetermined, as do house-lot limits. The figures given assume a geometric symmetry and centered placement of the structural core that is not in fact supported by present-day ethnographic data (see Hayden and Cannon 1983, 1984).

12. Data from transects C and D are not presented due to their considerable complexity. This complexity arises not from any particularly unusual or anomalous conditions present in the sampled areas but from the combined effects of temperature-induced distortions involving the transect C and D lab results (see n. 1 above) and an inadvertent displacement/juxtapositioning of several sampling loci in the neighborhood of the two transects' conjunction. In fact, the two reflect situations generally comparable to those indicated by our other tests: expanses of slightly to moderately elevated phosphate levels extending between the mound groups with continuous zones of significant enhancement concentrated within the fourteen meters or less immediately adjoining each structural core.

13. The "buried" structure dates from the Middle Classic period (ca. A.D. 550–650); the others date from after A.D. 700.

14. In SAR terminology, groups 2E-2 and 2E/F-1 are *plazuela* groups; 2C/D-1 and 3E-1 are patio groups. The *plazuela* groups pertain to our urban middle status category and the patio groups to our urban commoner one.

PART IV B. L. TURNER II & WILLIAM T. SANDERS

Summary and Critique

(*Editor's note:* The following comments by Turner and Sanders are revised versions of the commentaries made by the discussants at the original "Gardens of Prehistory" symposium in Toronto. Symposium participants have had the opportunity to revise their own papers as they have seen fit in light of these comments; however, differences of opinion regarding method and interpretation remain. The volume authors have attempted, and for the most part have been successful in their attempts, to adhere to the terminology suggested by Turner below. The discussants' comments provide both a general and a specific critique of the papers as they now stand as well as a valuable perspective on the research process within the study of prehistoric agricultural systems in Mesoamerica and adjacent regions. A full understanding of some of the more detailed comments contained herein will require a careful reading of the chapters referred to and the reader's own assessment of the degree to which individual authors have addressed the comments provided by Turner and Sanders.)

COMMENTS BY B. L. TURNER II

Some trends within Mesoamerican-Southwest studies—the regional focuses for the chapters in *Gardens of Prehistory*—seem symptomatic of larger undercurrents within archaeology and the study of prehistory in general. These fields are marked by philosophical pluralism in which multiple paradigms are competing for dominance, if only momentarily. This pluralism even extends to the next conceptual tier, that of theories of general processes. In such times, the apparent impossibility of achieving agreement on broader issues frequently shifts research attention to more meso- or micro-level themes and issues and to the application of new or revitalized means of data retrieval and assessment, as embodied in many of the works in this volume.

It may be, on the other hand, that this shift in emphasis is the cause of the pluralism, not its consequence: the increasing systematization of scientific information into ever more refined and confined bodies of knowledge has the effect of isolating specialists and providing them with a sufficient audience that their interests (and expertise) remain those largely internal to the specialty. This situation can lead to problems of synthesis—of making specialized data meaningful to larger and more integrative problems.

Such trends make archaeology particularly susceptible to the problems of "specialist-synthesis" dichotomy described for geography (Turner 1989) in which the required level of expertise in subfields and techniques apparently leads to personal research trajectories that are at odds with the disciplinary demands for synthesis. While the specialist (e.g., ceramacist, pedologist, palynologist) who produces evidence that potentially has bearing on a wide range of subjects and themes (e.g., agriculture, social organization, migration) is expected to demonstrate that significance, serious problems may be encountered because good synthesis requires more than a rudimentary understanding of the subjects to which the specialist is linking. Even research in the more synthetic segments of the field, such as settlement studies, can suffer this problem if care is not taken to entertain thoroughly the evidence from outside the region/locale and the subject beyond the confines of the home discipline.

Introductory Comments

This collection of studies, taken as a whole, tackles the subject of ancient agriculture in the Americas, especially its internal spatial form from the house-lot garden to more distant fields, with a refreshing problem-solving attitude of "getting on with it," without genuflecting to perceived wisdom or an overarching theory. The approaches are empirically based and demonstrate how special field and analytical techniques can provide further evidence of the kinds of agricultural situations that may have been present in the past. For the most part, specific problems of technique and analysis are best left to the specialist; problems of synthesis, however, are another matter. Here we struggle with, not against, the authors as they attempt to connect the new evidence to the larger subject in question, facing the obstacles of sheer complexity and the level of synthesis required. After all, agriculture, even if confined to "gardens," is the outcome of complex and multiple-level decisions made within the structure of environmental, demographic, and socioeconomic variables, including influences from afar.

Three broad issues link these chapters as a group: identification/typologies, form/function, and agricultural landscape changes.

Identification/Typologies. Students of New World prehistory (not archaeologists alone) have created their fair share of ad hoc agricultural typologies and terminologies, including plot types, as if indifferent to those that exist in the larger field of agricultural systems. This discordance not only inhibits comparative studies but can result in erroneous identifications of agricultural features and measures. If comparisons are important—and they are for the development of useful form-function models such as Killion's house-lot model—then standardized meanings and classifications must be employed. To this end, a number of terms used repeatedly in this volume have rather specific and universal meanings in agricultural studies.

GARDENS (kitchen, house, or compound) *are spaces for the cultivation of multiple species used for additional or emergency caloric and nutritional needs, medicinals, ornamentals, and other "exotic" production.* Evans and Doolittle both provide evidence for the use of gardens as an emergency source for staple seeds and foods. Special varieties of staples for ceremonial and other such activities may also be grown there. Almost invariably gardens are located immediately adjacent to the farming abode or within the perimeter of the farming compound, are small in size (especially compared to fields devoted to staple foods), have high labor and other inputs, and commonly are the domain of women, often involving codified proprietary rights.

Gardens are universal to almost every type of agricultural system, both past and present, regardless of environmental conditions, the intensity of the system for land economy, or other such factors. Gardens are not solely a tropical phenomenon. Cultivators in the tropical world may, on the average, derive more output from their gardens than those in other climatic conditions, but land pressures and other factors that raise or lower the significance of gardens to the total system of cultivation must also be considered. Claims about the special significance of gardens in the tropics may result from confusing the term *garden* with *horticulture* (see comments below).

ORCHARDS *constitute space dominated by economic tree or treelike species, usually domesticates.* Such plots are situated at varying distances from the farming abode, although locations near the house seem favored. Typically, one or only a few species are cultivated. The extensive use of true orchards, as opposed to areas culled for useful species, may not have occurred throughout most of the preconquest Americas.[1] The analogies used by Maxwell and Anschuetz for the Southwest are ones that were highly influenced by the Spanish. Obviously, additional research is needed on this important topic.

ORCHARD-GARDENS *are spaces in which the functions and characteristics of a garden are mixed with those of an orchard.* This type of plot is typically adjacent to the farming abode or within the farming compound and may have distinctive areas for trees and other cultigens. Perhaps more common to the tropics, orchard-gardens may have been used throughout much of Central America and Mexico in ancient times.

INFIELDS and OUTFIELDS *constitute spaces cultivated primarily with the major staple crops, the former closer to the main farming abode than the latter.* Apparently first employed in descriptions of field systems in Europe (see Grigg 1974:62), the two terms were developed to describe the internal farm or village arrangement of fields in rain-fed agriculture, such that one cannot exist without the other; if both are not present, then the prefixes to *fields* would have no meaning. Both fields may be intercropped or cultigens may be rotated. Owing to locational advantages, the infield is usually cultivated more frequently than the outfield, receiving more inputs and having a higher output per unit area and time. The outfield may be so extensive as to qualify as a swidden. I am unaware of comparative studies of the distances from field to abode, although this measure is likely to vary by context.

In most instances, the chapters in this volume employ these terms correctly in terms of their functional definitions. Some problems are inevitably encountered, however, in the effort to insert location (plot to abode distance) into the basic definitions (e.g., Killion and Santley), a subject discussed below. In others, potential confusion is encountered through the implication that distant fields are gardens (e.g., as used by Maxwell and Anschuetz) and that fairly distant fields (2.68 kilometers from residence) are infields (e.g., Santley). In yet others, it is difficult to determine if the authors are suggesting the past existence of gardens, orchard-gardens, infields, or all three (e.g., Ball and Kelsay). And, of course, the existence or significance of outfields is totally inferential (e.g., Zier). Additional research is clearly necessary in order to fine-tune these terms, given the problematical nature of the archaeological record.

HORTICULTURE *literally means the cultivation of a garden or orchard,* but it has taken on a different meaning in the anthropological literature on agriculture and land use. Here it is commonly applied to hoe-based, polycultural cultivation, typically emphasizing root crops and tubers, apparently in an attempt to distinguish this system from *agriculture,* or a plow-based, monocropping system of cereals. These distinctions reflect parochial biases and create more problems than they solve, including the implication that the anthropological "horticultural" plot is a garden, when in fact it is an agricultural field by function.

It must also be noted that a multitude of field systems, including rain-fed ones, do not fall into the infield-outfield types, as illustrated for the U.S. Southwest by Maxwell and Anschuetz and for the Basin of Mexico by Evans. Most intensive land-use systems, for example, generally do not have outfields because that space is occupied by other settlements and farms. Moreover, agriculture dominated by irrigation, wetland, or terrace systems typically have no identifiable outfields.[2] Indeed, Evans's discussion of the *calmil* suggests that the garden and infield can be fused in parts of Mexico. My observations of them suggest that in many *calmilli* the boundary between the garden proper and the infield is not formally demarcated, although it can be detected by the change in cultigens.[3]

Specific Comments

Form and Function. The desire to fit plot forms (or sets thereof) to their typological functions is understandable for the student of prehistory, because attributes of form can be detected in the archaeological record while function can only be inferred. Geography's long search for links between spatial form (distance and location) and function has delivered several important lessons here: general relationships hold only ephemerally because context is so important (see Sack 1972; Szymanski and Agnew 1981), and the validity of the link can be improved only by increasing its specificity (and hence lessening its general applicability).

Throughout this volume two aspects of settlement form have been used to infer possible agricultural plot function: the cleared, "rubbled," and cultivated spaces of the area adjacent to the farming abode as evidence of gardens and infields, and the density of "rural" settlement as evidence of the small space for garden-infields and, hence, the apparent need for outfields. These interpretations are valid possibilities but not the only ones. The chapters by Doolittle, Evans, and Maxwell and Anschuetz indicate some of the range of possibilities in plot systems. Expanding the analysis of this range would reveal a greater array of "forms," many of which have similar spatial patterns but have dissimilar agricultural functions.[4] Maxwell and Anschuetz allude to this problem in reference to relict fields located away from the settlement but with forms similar to those found in the gardens adjacent to the settlement. Evans's study of rural Cihuatecpan cannot detect garden from infield as per the *calmil* issue noted above. Santley recognizes the possibility of "inlying" and "outlying" infields. And Killion's (1987a:figs. 32–33) detailed study of contemporary form in the Tuxtlas region demonstrates a high percentage of infield-outfield reversals in terms of the intensity of cultivation, such cases invariably the product

SUMMARY AND CRITIQUE

of either extreme differences in land quality or intense land pressures. Such "exceptions" or deviations from more common patterns should not derail our search for commonalities in form-function. They do, however, remind us that this match requires a "best fit" approach, following Evans, based on the general agricultural patterns, the totality of prehistoric evidence, and a consideration of the effect of expanding the spatial scale of analysis.

Zier, Santley, Killion, and Ball and Kelsay each make a good case for garden-infields and outfields at the sites in question because of identifications made with the house-lot zone and the apparent small size of these plots for producing sufficient food to meet subsistence needs. The approaches used to infer garden-infields are impressive—cleared and "rubbled" spaces, phosphate loadings, buried ridges and furrows, and so on. Intensive cultivation in the vicinity of the farming abode was undoubtedly practiced at each of the sites in question. The evidence for outfields, however, is solely inferential. Santley and Killion seem to rely on the house-lot model and make no argument about the production capacity of the garden-"infield" and the demands placed upon it (e.g., population density). The two other chapters briefly make this case but based on assumptions, either implicit (Ball and Kelsay) or explicit (Zier), about land productivity (and yield) that are drawn from other, "undefined" systems of cultivation. Their appropriateness, therefore, is difficult to assess. It is noteworthy that many of the measures of land productivity used for analogies in Mesoamerican studies are drawn from extensively cultivated plots and from farming systems that are supplementing food production through wage labor, not from intensively cultivated subsistence systems.

Outfields are most difficult to identify archaeologically because their very extensiveness generally precluded permanent land changes that can be recovered, and if these were made, it is a strong signal that the field in question was not of the outfield kind. Presumably, the case for outfields must be made on arable lands in the vicinity of the site that are "vacant" in the sense that no other land-intensive activity can be documented there. To demonstrate this condition requires that the spatial scale of the archaeological investigation/interpretation be enlarged, a task that most of these studies did not address. Interestingly,

McAnany does expand the scale for the sites around Pulltrouser Swamp, demonstrating that much of the possible land for outfields falls into the territories for other sites.

In the Pulltrouser example, the problem of outfield identification is further heightened by the locations of the swamp-rim settlements relative to their fields. Many of the raised-field complexes within the swamp proper lay at distances that are equivalent to the distances of dryland fields. Given that the low ridge to the west of Pulltrouser Swamp—the most archaeologically "vacant" zone—is dominated by rich mollisols and vertisols that are prime agricultural soils in which the major impediment to cultivation is insufficient moisture during the dry season, I suspect that no true outfields existed.[5] Fields within the swamps and on the ridge were cultivated intensively, at least in terms of the frequency of cropping. The agricultural form on the west side of the swamp, then, is one of spaces around farming abodes (gardens?), wetland beds within the swamp, a small open zone on the low ridge (infields?) that also falls into the potential infield area of other sites, and no zone for outfields because the larger area was thoroughly settled. Omitting the wetland fields, the form suggests gardens and intensively cultivated "near-fields." This same form is found at each of the other sites mentioned above and, therefore, the same functions might also have applied, especially at Guerra (Ball and Kelsay) where alluvial soils ring the site and limestone-based soils occur outside the river valley proper.

Finally, I would like to comment on a number of references in the volume on the relationship of form-function to the environment. Environmental influences on form-function are undeniably important, but, as in other comparative studies of agriculture (e.g., Turner, Hanham, and Portararo 1977), this influence is probably not primary but shares a complex role with conditions of land pressures, cultigens, and cultural tradition, among others. In high population density conditions in the wet tropics, for example in southern and eastern Asia and highland eastern Africa, well-defined and demarcated boundaries for various agricultural activities can be found within the individual farming complex and the village, and these do not necessarily follow the house-lot model. Again, the point is one of complexity and of the need

SUMMARY AND CRITIQUE

to look beyond the relatively sparsely settled conditions of the Mexican and Central American tropical lowlands.

Agricultural Landscape Changes and Intensification. Unless otherwise defined, *agricultural intensification refers to increases in inputs (including labor) and output (yield and total production) per unit area and time.* Typically, the relationship between the two variables is strong and positive, and this same relationship can be found between the two variables and the frequency of cultivation or the amount and kind of landscape capital (see Turner and Brush 1987), hence Boserup's (1965) use of the frequency of cultivation as a surrogate measure of intensification. Environmental conditions, of course, are a powerful mitigating variable in these relationships, offering different levels of opportunities and constraints to intensification and, hence, affecting the nature of landscape capital employed.

Landscape capital, following Blaikie and Brookfield (1986), *refers to modifications of the landscape or field/plot for the purpose of sustaining or improving cultivation.* While any number of classifications of these modifications can be devised (for Mexico and Central America see Wilken 1987), the major classes include slope modification (terracing), movement of water to fields (irrigation), removal of water (drainage), and multifunctional earth or field raising. The last is very important to all kinds of agriculture and was used extensively in the New World (Denevan 1970); but it has been one of the more misunderstood forms of cultivation among students of New World prehistory, and the nomenclature applied has caused confusion.

Denevan and Turner (1974) have used the generic term "raised fields" for all instances of piling or raising earth, be it on slope, in wetlands, or as mounds, ridges, or rectilinear beds. This generic term, unfortunately, has taken on the implied meaning of wetland fields in much of the Mesoamerican literature. While Zier is not wrong in labeling the ridge and furrows found at Cerén as raised fields, the term *ridge and furrow* is more precise because it identifies a type of raised field common to rain-fed, dryland agriculture.

A comment is in order about the function of field raising and its purported association with marginal agricultural lands. The technique in varied forms is used on all types of lands and soils with great ranges in native fertility for cultivation; it is not the product solely of cultivation of "marginal" lands or soils of low fertility. Denevan and Turner (1974) have detailed some of the large number of production-enhancing functions of field raising.[6] These, of course, vary by environmental conditions and cultigens employed. In Mesoamerica and elsewhere in the prehistoric New World, rain-fed maize was almost universally cultivated on some form of "raised fields," be they in the form of ridges or mounds *(camellones, montones)*. Apparently this practice was even used in extensive fallow systems, although the elaboration of the features may have diminished in these cases. It must also be recognized that the more elaborate forms, such as the *chinampas* or other wetland systems, can elevate "marginal" lands (typically marginal only in the sense of waterlogging) to a status that is superior to that of the former "prime" drylands.

Finally, some comments concerning McAnany's notion that intensification in the prehistoric and historic wet tropics did not involve a shift in tool technology as did that in the more temperate environments. This is an interesting idea, but it requires considerable elaboration and reflection. All agricultural land intensification has as a goal the increased or sustained enrichment of the soil; this goal cannot be ascribed solely to the tropics. Intensification in temperate climates occurred long before the plow was invented, so that the pre-plow intensification-tool relationship demands attention. Moreover, the adoption of the plow has as much or more to do with labor issues than with improved cultivation conditions. A common myth, for example, is that grasslands could not be cultivated without the plow; this is not so, but without the plow it was a most labor-consuming activity. My guess is that the tool-intensification association has much more to do with the development of metallurgy and changes in cultigens than it did with broad environmental zones.

Odds and Ends. The lack of plant casts in the ash-covered ridge and furrows found at Matacapan (Santley) can also be ascribed to the deposition of the ash before planting; a cleared or harvested field of maize almost always retains stalk and debris to be turned into the soil for organic nutrients. Likewise, the single casts of stalks found at Cerén might well be interpreted as transplanted maize—and hence a most intensive system—rather than a procedure against infertile soils as Zier suggests.

It would be helpful if Ball and Kelsay would infer more about the apparent lack of garden-infields at Buenavista and the meaning of this at Guerra and in the general area. Was Buenavista supported by sites like Guerra, and if so, what does this relationship mean for the intensity of cultivation? Also, given the general uniformity of soils at Guerra, the interpretations of the locations of possible agricultural plots there seem to be quite sound.

Finally, realizing that this topic exceeds my expertise, I wish to comment on the tools at Pulltrouser Swamp. First, an intact tranchet biface pick was found in a wetland there (Shafer 1983:fig. 12-3j); but this field was a channelized type, one in which ditches or channels were apparently dug from the swamp to the mainland proper (Turner 1983b) and one that may date as early as the Preclassic period. No complete lithic was taken from a field in the swamp proper. Second, the apparent simplicity of tools in the Maya lowlands is interesting. Ethnohistoric accounts describe a footplowlike instrument used in the lowland region (Herrera y Tordesillas 1726–27:133–34), and the digging stick or dibble as possibly depicted in Maya codices indicates a waist-high instrument with a curved upper portion and either a sharp or blunt point, reminiscent of the Peruvian *huiri* or *taclla* (footspade) used to turn earth.[7] I am also struck by the artists' renditions of Amerind tools used for soil preparation (albeit with artistic license), some of which depict hatchet-shaped instruments (see Weatherwax 1954). Finally, I suggest that the massive clays in Pulltrouser Swamp would require a footspade or hefty stone instrument for digging.

COMMENTS BY WILLIAM T. SANDERS

A major problem in documenting the prehistory and evolution of agricultural systems, whether the objective is that of house-lot gardens or any other component of the agricultural system, is of course methodological. In general, our conclusions are based on indirect data because we cannot actually observe the behavior of prehistoric subsistence cultivators. In such reconstructions the use of ethnographic analogy is common. While I believe ethnographic analogy is a powerful tool and virtually all writers in this volume use it, the most important contribution of the book is that it applies a variety of methodological approaches and techniques to test ethnographic models formally. These techniques and methods include surface survey, the oldest and most commonly used method, but also include test excavations, pollen analysis, soil chemistry (phosphate) testing, and lithic analysis.

The chapters by Killion, Santley, Ball and Kelsay, and Evans all use highly structured settlement pattern surveys and spatial analysis of dwellings to test ethnographically derived models of land use. These activities are also accompanied by a program of test excavations. The major advantage of this approach is that data are produced for very large areas. The disadvantage is that there is probably more than one single model of land use that may be fitted to the survey data, even when coupled with test pitting. By using only these two techniques it is often impossible to discriminate among these alternatives.

Killion has presented an excellent ethnographic model, based on present-day patterns in the Tuxtlas region of southern Veracruz, Mexico. The model he produces is an appropriate one for much of twentieth-century Mesoamerica. It is found in areas where nucleated settlements, usually of relatively low density, are found and involves a pattern of residences within the village itself that includes the following components: (1) the actual residential buildings with several different but overlapping functions, (2) a nearby cleared space used for work activities, and (3) a surrounding household garden that makes up most of the surface area of the house lot. What makes this model useful to archaeologists is that he records the activities that take place in each area and observes the refuse distribution from area to area. Areas 1 and 2, for example, are kept relatively clear of refuse and the garden has a moderate to low refuse concentration. The densest refuse is found in a narrow transitional zone (4) between the work space and the garden. He goes further, however, and shows how the distance between residence and agricultural fields outside the village affects the uses of different components of household space within the village. He finds significant correlations between the size of infields and outfields and the size of noncultivated parts of the house lot.

SUMMARY AND CRITIQUE

It should be noted here that Killion uses the term infield very differently from the use intended by Palerm. Palerm's infield was the house-lot garden or *calmil*. Because significant variability exists in the Tuxtlas region, in terms of the distances between houses and agricultural fields (what would be called outfields by Palerm), and because these differences have an impact on the use of space within the house lot, Killion essentially divided Palerm's outfield into two subtypes, fields lying within a forty-minute walk of the house, which he calls infields (usually cropped more intensively), and those beyond this distance, which he refers to as outfields. I should note here that in the more densely settled highlands of Mesoamerica today, particularly in semiarid regions, all agricultural fields are intensively cropped, and all village lands are within a forty-minute walk, so in these areas this distinction is not relevant.

The specifics of his model then should be applicable, with restraint, to archaeological contexts. Killion applies the model to the Middle Classic settlement pattern data from the same area rather successfully. Santley's attempt to apply it to the Middle Formative is an interesting preliminary effort but, I would argue, because of the much poorer data base needs much more testing.

I see three major problems in terms of the usefulness of the model when applied to other areas. First, even assuming that all of the prehistoric occupation is revealed by a survey, combined with test pitting, the data are usually coarse grained in temporal terms. In most projects, for example, remains that are dated to a single ceramic phase are considered contemporaneous. These phases, however, are often two to three centuries in length. Based on ethnographic cases, all residences would never peak in population simultaneously; furthermore, older residential sites would frequently be abandoned and new ones formed during a period this long. What this temporal pattern means is that the surface survey recovers an aggregation of a series of superimposed settlement systems, whereas the model, to work, must catch the system in a moment of time, ideally a single year.

Second, in cases where houses are built on the summit of substructures and/or the superstructures are built of relatively massive adobe or masonry walls, much refuse is incorporated in the fill. Because of this factor, in central Mexico the densest refuse on a site is usually directly

275

associated with buildings themselves, and in Killion's model this should be an area of very light refuse. When such sites become eroded, what ensues is a pattern of gradually decreasing debris from the preserved building cores, fanning outward to form a nearly continuous area of heavy to moderate density and then tailing off to lighter and sparser occupation on the peripheries. His model would then work best in areas where the dwellings were constructed of perishable materials.

The third problem I see is that in most Mesoamerican sites, regardless of the environmental context, the household was a relatively large extended family, and the physical remains consisted of a series of buildings, residences, kitchens, and storage structures, often arranged around a patio. The patio was used as the work space and corresponds to Killion's clear area in terms of refuse density and function. Buildings, as in his model, tend to be relatively free of refuse. Zone 3, the band of dense debris, in patio groups usually occurs immediately behind the structures and against the back retaining walls of their substructures. This is the typical pattern at Copan and other Classic Maya sites.

The chapter by Ball and Kelsay applies the method of phosphate soil analysis to determine variety in the use of a prehistoric landscape in Belize. Ball and Kelsay combine the method with surface survey and test excavations over a relatively large region, characterized by abundant and substantial surface remains of houses. This aspect of their project is comparable to the research of Evans and Killion.

My overall reaction to the study by Ball and Kelsay is that phosphate and other chemical analyses of soils have the potential of becoming powerful analytical tools in the study of prehistoric land-use patterns. The problem I see is how to factor out what I call background noise, that is, the effect of recent land use on the soil characteristics. I am not yet convinced that the problem can be easily resolved or at least is feasible in terms of time and financial resources. I also see the same problems I noted for the surface survey/spatial analysis approach. What we are detecting in the soil analysis is a composite picture of numerous settlement systems superimposed over time. The question is first, how many samples do we have to take before we can say that a pattern of distribution exists that is regular rather than irregular? And, more im-

SUMMARY AND CRITIQUE

portant, if regular, how does one sort out the effects of multiple agricultural occupations? An additional problem is how does one procure a reference soil sample as a standard—that has never been affected by human occupations—from an environment used by humans over thousands of years? I see this as a serious problem, particularly in the Belize study, for it would be difficult to find any area near their sites, which was cultivable, that had not been cultivated at some time in the past. The problem of reference standards then is crucial.

Ball and Kelsay apply phosphate analysis to a settlement system in Late Classic times consisting of a dispersed, dense rural settlement. The method in the Belize case worked very well in detecting buried residential sites, all, interestingly, dating prior to the Late Classic surface remains. I am less convinced that the lower- or intermediate-level samples, however, necessarily represent evidence of prehistoric land use. I am particularly puzzled by the "off-site" samples. The overall population density of the region was so high that it is very improbable that any off-site land (meaning outside the actual residential site) was unused, and I would argue that it must have been used intensively. Why then should there be less phosphate in the vacant terrain between house clusters? The only distinction should be between house-lot or kitchen gardens, where refuse from the household would be heavy in contrast to all other fields. Finally, I suspect that some of the variety seen in their samples is due to differences in natural soil fertility as well as human use.

The above remarks should in no way be understood as a criticism of the research by Ball and his associates. I think it is an exciting project and the combination of methods used here has great potential in the recovery of data on prehistoric land-use patterns.

Zier's chapter is an interesting example of how far an archaeologist can go when faced with a Pompeiilike situation in which a very specific archaeological context is frozen in time. He uses a variety of techniques to analyze the meaning of that context: lateral exposure of the site, including a house and adjacent garden sealed off by an ash fall, deep coring, pollen analysis, resistivity, radar, and other techniques of remote sensing.

The major conclusion of this research—that it reveals a pattern of residences, infields, and probable outfields—I find very convincing. I would, however, make two suggestions. First, the sequence of ridges and furrows Zier describes is not a distinctive *type* of cultivation; rather, it represents a phase of the cultivation cycle and is found all over highland and lowland Mesoamerica where fields are intensively cultivated, at least intensively enough to require manipulation of the soil surface. Features like he has described have been found in other sealed contexts, under architecture at Kaminaljuyu and Matacapan and under ash fields at Huexotzingo in Puebla. In highland Mesoamerica, in the late summer to early fall, hoes or plows are used to heap soil around the bases of maize plants to form either hills or continuous embankments. The purpose is a bit obscure; most farmers say it helps protect the crops against wind damage, and others say it adds fertility and conserves moisture. In highland Guatemala the technique is used to produce a pattern of hoe contouring on slopes to control downslope drainage. What the presence of these features in Zier's site suggests is that the volcanic ash flow occurred sometime between early August and late September and happened to cover the field at this phase of field preparation.

A second suggestion has to do with the shallow depth of the white tephra, its low fertility as well as its water retention quality, and the potential use of the red clay below it. In the more arid regions of highland Mesoamerica today a technique of planting is used called *cajete* or *a todo costo*. Prior to the inception of the rainy season, and with the purpose of getting a head start, a series of shallow pits is excavated down to the humid subsoil, maize seeds are placed in the humid soil at the bottom of the pit, and the pit is then partially filled with dry soil mulch. The technique guarantees an early start on the rainy season. As the rainy season proceeds, the pit is generally filled in. In such fields, then, the topography shifts from a field full of shallow depressions to one of banks and furrows. At Cerén the added value of the method would be to tap the fertility of the underlying red clays, but it would also help to resolve the problem of the poor water retention qualities of the tephra.

The chapter by McAnany approaches the problem of reconstruction of prehistoric agricultural habitats from an entirely different perspective: the use, refurbishing, and ultimate discard of basic agricultural tools—an intriguing methodological approach, especially when com-

SUMMARY AND CRITIQUE

bined with regional survey, as in her case example, and with soil chemistry studies, apparently not yet conducted at Pulltrouser. I have some reservations and disagreements with several of her statements, but as a whole the chapter offers a convincing reconstruction, particularly of the Late Classic–Terminal Classic agricultural system. I will summarize first my areas of agreement.

First, the overall Late Classic–Terminal Classic settlement pattern of the Pulltrouser Swamp sites (perhaps better labeled swamp-margin sites, as they are all located in well-drained terrain, not in the swamp, and in this respect the pattern is similar to Classic Maya sites everywhere in the southern Maya lowlands; see my extended comment below) consists of the typical Classic Maya dense but dispersed rural settlement, distributed on ridges, hills, and flatland on the swamp margin. The density of remains clearly indicates some form of intensive cultivation. Whether one calls the adjacent fields house-lot gardens or infields, in this case, is an academic point. The space between each house measures between .5 and 1 hectare in size and thus represents sufficient cultivation space to supply a small extended family with all of its food requirements. Those areas closest to the residences would receive nutrient subsidies from household refuse, the more peripheral spaces very little. The model also fits the distribution of chert tool debitage and discard that is the meat of McAnany's chapter.

Second, the bifacial tools she describes are undoubtedly tools fabricated to work and turn over the soil and are an indication of some intensive cultivation; both the shape and the wear patterns strongly suggest this conclusion.

My disagreements with her conclusions are as follows.

First, considering the dense spacing of nearby unsurveyed sites, at distances three to four kilometers on the map (these distances must refer to the centers of such sites because the dispersed settlement pattern probably looked continuous over the whole region of her map; see McAnany, this volume, fig. 8-4), leaves little or no room for outfields in Late Classic–Terminal Classic times. In my opinion, all of the agricultural land used by the population of the Pulltrouser settlement system was within the two- or three-kilometer radius shown on her map. I suspect (and this was Netting's [1977] model of Classic Maya land use) that the southern Maya heartland—that is, much of central and northern Belize, the Peten, Campeche, and Quintana Roo—was virtually a continuous zone of infield cultivation.

Second, I would suggest that the data from Colha, the source of the heavy chert tools for much of central and northern Belize, including apparently Pulltrouser, indicate a shift from emphasis in chert tool manufacture from axes in the Preclassic period to hoes in the Classic. Shafer (1983) makes the conclusion in his study of the Pulltrouser Swamp lithic assemblage. He suggests that thicker, more massive, tranchet-bit tools were used as axes and the oval bifaces as hoes. Notice he uses the term *hoe* for the latter. McAnany believes that the bifaces could have been hafted like celts, based on one with the haft still in position found by Puleston in a canal near a raised field at another raised-field site in Belize. McAnany presents a series of ethnographic cases where weeds are removed by slashing at the base, sometimes penetrating the soil and even cutting the weeds off at the roots, and suggests that the bifaces could have been used in this manner. The tools used in her ethnographic cases, however, are all made of steel or iron and are considerably thinner than the chert bifaces; the latter would be a very clumsy and ineffective slashing tool. Her bifaces are very similar to tools described in the literature as hoes in both the Old and the New World. Hafted as a hoe, they would be much more effective as weed removal and soil preparation tools.

A third point is that the sample size used in the analysis of phases prior to the Terminal Classic will have to be increased in the future in order to test more fully her model of increasing intensiveness of land use. The process that McAnany discusses makes sense; what I am challenging is the evidential aspect.

Fourth, I also would challenge the notion that the raised fields were a significant component in the economy of the settlements from which McAnany's sample derives. While she claims that bifaces were, on the basis of wear patterns, clearly used to work well-mulched soils, such as those found in drained fields, one can mulch agricultural fields regardless of their location, including well-drained slopes. Evidence will have to be recovered in the future that documents more clearly that these tools were in fact used in the cultivation of such fields. Operations 3 and 4 at Pulltrouser, areas of drained-field testing, yielded only 17

SUMMARY AND CRITIQUE

tool fragments and 54 pieces of debitage, in contrast to the 280 and 1,536 artifacts found in excavations of one of the residential sites on well-drained terrain. Virtually all of the data presented refer to intensive cropping of well-drained fields during the Terminal Classic period. The question must be asked then, accepting the small sample from pre–Terminal Classic times, and given her model of increasing intensification of agricultural land use, why would the Maya have cultivated the swamps as early as the Late Preclassic phase?

The chapters by Doolittle on the "Gran Chichimeca," the desert region of northern Mexico, and by Maxwell and Anschuetz on the Chama Valley of New Mexico both deal with settlement agriculture in a very arid environment north of Mesoamerica.

The first part of Doolittle's chapter consists of a discussion of kitchen-garden agriculture among present-day Pima, Tarahumara, and Tepehuan Indians in Sonora. These are all heavily fertilized irrigated plots, of diminutive size, on which a great variety of crops are grown. Residences in the ethnographic cases seem to be on high ground with the gardens placed immediately downslope, where they can receive drainage and refuse from the nearby higher terrain. They are also found primarily on the sunny sides of slopes. One very interesting point that Doolittle makes is that some maize is grown in these diminutive gardens, an apparent anomaly inasmuch as much larger outfields of maize are cultivated and provide most of the calories. He suggests that the growing of some maize in kitchen gardens ensures the production of seed for the next year's planting, an important need, considering the risks of crop production in this arid environment.

He then applies the kitchen-garden model to the prehistoric remains in the same area. Surveys have detected a great variety of features that are probably agricultural in function: tiny circular walled spaces, much larger rock enclosures, parallel rock walls constructed along slope contours, and true terraces. These are all found downslope and located just below nearby house sites. They seem to occur particularly in relationship to settlements that he calls *rancherias,* that is, small rather dispersed settlements usually placed on tops of mesas. Anasazi-like apartment-type villages also are found and here the association of plots to residences is less clear. While no direct evidence exists of function,

Doolittle feels that the small enclosures were seed beds and the rest probably functioned as household gardens. It is the analog with present-day practices that strengthens his conclusions.

The Chama Valley study by Maxwell and Anschuetz reveals a pattern even further removed from the house lot–kitchen garden/infield/outfield trichotomy as one moves into the colder and more arid environments of North America. Here settlements consist of large dense apartment house villages around which a great variety of fields are cultivated within a few kilometers of the village. Variation among them represents an adaptation to the key problem in this area—water availability. Because of high internal residential density, no space is available within the settlement itself. On the basis of ethnographic analogy, drawn from the same region, the authors make a distinction between gardens or small plots containing a great variety of crops, usually planted close to the village where they can be easily tended (primarily by women), and outlying fields dedicated to the production of staple crops like maize. The gardens are extraordinarily small plots; in one case 113 cultivators cropped only fourteen acres of land. The outlying fields are cultivated, to ensure adequate moisture, using a variety of techniques such as dry farming, terracing, floodwater farming, or permanent canal irrigation; or they are placed in locations that receive natural moisture through underground seepage, the latter occurring in a variety of contexts such as floodwater plains, outwashes, and areas of unusually high water table. Two additional techniques, found only in archaeological contexts, include the construction of low parallel walls along slopes to trap drainage, approximating therefore terraces in function, and gravel mulching. The Southwest is one of those areas where much of the activity of agriculture is dedicated to resolving the twin problems of water availability and temperature, and both ethnography and archaeology provide us with a picture of hard-pressed agricultural populations using a great variety of ingenious technologies to ensure crop production. In summary, what was grown, where and how, is not easily resolved into an orderly pattern of gardens, infields, and outfields, and this more complex set of spatial arrangements and their archaeological consequences are well documented in the study by Maxwell and Anschuetz.

Variation in the spatial patterning of field types and the agricultural

SUMMARY AND CRITIQUE

practices employed on them are to a great extent due to the combination of demographic and environmental conditions faced by subsistence cultivators. The Mesoamerican pattern of garden, infield, and outfield is particularly well adapted to tropical areas such as the humid lowlands, where the overriding problem of agricultural intensification is that of soil fertility maintenance. In the semiarid Mesoamerican highlands fertility still can be a concern, but in limited areas the practice of irrigation, particularly floodwater irrigation, indirectly resolves this problem and permits a wider expansion of stable intensive agriculture. In the more arid regions such as the Southwest the overriding problem is not fertility but water and a distinctive set of implications for the spatial organization of land use and agricultural production.

NOTES

1. Orchardlike arrangements have been suggested for the production of cacao (*Theobroma* spp.) in Central America (Wilken 1987), and ethnohistoric accounts (e.g., Oviedo y Valdes 1853:230) suggest large areas devoted to tree crops in the Yucatan area, although the latter may refer to multiple orchard gardens adjacent to houses.

2. Doolittle (1984c) has described the ephemeral cultivation of gullies and washes in the Valley of Sonora, a practice that might be interpreted as outfields, compared to intensive cropping on the valley floor. Interestingly, in many cases the gullies are located closer to the ancient settlements than is the valley bottom, and continued cultivation of a gully can transform it into an infieldlike plot.

3. Note that the term *milpa* literally means maize field or a field dominated by *Zea mays*. It does not mean outfield or swidden. Evans is correct to note that it implies little about the intensity of cultivation or technology used. A *milpa* in the sparsely settled humid tropical regions of Mexico and Central America may have characteristics akin to horticulture, while those in more densely settled or xeric regions may be monocropped.

4. The form-function relationship in agriculture is not merely an environmental association but involves cultigens, population pressures, and socioeconomic conditions, among others (see Sanders and Killion, this volume). For examples of the range of form-functions see Turner and Brush (1987).

5. Northern Belize is dominated by limestone-based mollisols and vertisols

of high agricultural fertility by almost any standard. Long linear outwashes of granitic-based soils overlay portions of the region. Soil quality, then, is directly related to the presence of the dark brown, limestone-based soils. Almost all of the area surrounding Pulltrouser Swamp displays the latter; indeed, the entire western side of the swamp has been subjected to continuous sugarcane cultivation for well over two decades.

6. David Thurston and his students in the Department of Plant Pathology, Cornell University, are documenting yet another function of field raising: its positive effects in protecting the roots of cultigens from plant pathogens.

7. I have listed some of the ethnohistoric descriptions of agricultural implements used in the Maya lowlands and interpreted some Maya paintings of "staffs" as illustrative of them (Turner 1974:154–57). In that same effort I neglected to mention the large number of macroblades taken from excavations in terraced fields and terraces in the Rio Bec region.

References

Abrams, Elliot M.
    1988   Investigation of an Obsidian Midden at Cihuatecpan, Mexico. In *Excavations at Cihuatecpan, An Aztec Village in the Teotihuacan Valley*, edited by Susan T. Evans, pp. 235–38. Vanderbilt University Publications in Anthropology, no. 36. Nashville.

Adams, R. E. W., Walter E. Brown, Jr., and T. Patrick Culbert
    1981   Radar Mapping, Archaeology, and Ancient Maya Land Use. *Science* 213:1457–63.

Adams, R. E. W., T. Patrick Culbert, Walter E. Brown, Jr., Peter D. Harrison, and Laura J. Levi
    1990   Rebuttal to Pope and Dahlin. *Journal of Field Archaeology* 17:241–44.

Alcorn, Janis B.
    1984   *Huastec Maya Ethnobotany*. University of Texas Press, Austin.

Anderson, Edgar
    1952   *Plants, Man, and Life*. University of California Press, Berkeley and Los Angeles.
    1954   Reflections on Certain Honduran Gardens. *Landscape* 4:21–23.

Anghiera, Peter Martyr d'
    1912   *De Obre Novo: The Eight Decades of Peter Martyr d'Anghiera*. Translated from the Latin, with notes and Introduction, by Francis Agustus McNutt. 2 vols. G. P. Putnam and Sons, New York and London.

REFERENCES

Anschuetz, Kurt F.
1976 The Hopi and Their Maize: An Ethnobotanical Perspective. Senior honors thesis, Department of Anthropology, University of Michigan, Ann Arbor.

1984 Prehistoric Change in Tijeras Canyon. Master's thesis, Department of Anthropology, University of New Mexico, Albuquerque.

Anschuetz, Kurt F., and Timothy D. Maxwell
1987 Agricultural Intensification and Diversification in the Northern Rio Grande. Paper presented at the symposium, Agricultural Land-Use Variability in the Prehistoric Southwest, 52d Annual Meeting of the Society for American Archaeology, Toronto.

Anschuetz, Kurt F., Timothy D. Maxwell, and John A. Ware
1985 *Testing Report and Research Design for the Medanales North Project, Rio Arriba County, New Mexico: NMSHD Project No. F-052-1(15)*. Laboratory of Anthropology Notes, no. 347. Museum of New Mexico, Santa Fe.

Archivo General de la Nación
1603 Congregaciones Serie. Vol. 1, Expediente 48, Folios 30–34.

Armillas, Pedro
1969 *The Arid Frontier of Mexican Civilization*. Transactions of the New York Academy of Sciences, 2d ser., 31:697–704.

1971 Gardens in Swamps. *Science* 174:653–61.

Arnold, P. J. III
1986 Ceramic Production and the Archaeological Record: Some Questions and Considerations. *Haliksa'i* 5:57–73.

Ashmore, Wendy
1981a Some Issues of Method and Theory in Lowland Maya Settlement Archaeology. In *Lowland Maya Settlement Patterns,* edited by Wendy Ashmore, pp. 37–60. School of American Research and University of New Mexico Press, Albuquerque.

Ashmore, Wendy (editor)
1981b *Lowland Maya Settlement Patterns*. School of American Research and University of New Mexico Press, Albuquerque.

Avebury, J. L.
1869 *Prehistoric Times as Illustrated by Ancient Remains, and the Manners and Customs of Modern Savages*. Holt, New York.

Ball, Joseph W.
1983 Rural Community Structure in the Late Classic Maya Lowlands. Pro-

posal to the National Science Foundation. Ms. on file, National Science Foundation, Washington, D.C.

1991 Pottery, Potters, Palaces, and Polities: Some Socioeconomic and Political Implications of Late Classic Maya Ceramic Industries. In *The Peak of Lowland Maya Civilization: New Understandings of Eighth Century Maya Development*, edited by Jeremy A. Sabloff and John S. Henderson. Dumbarton Oaks, Washington, D.C. In press.

Ball, Joseph W., and Jennifer T. Taschek
1989 Secondary Centers and Classic Maya Political Organization: The Mopan-Macal Triangle Project. Paper presented at the 54th Annual Meeting of the Society for American Archaeology, Atlanta.

Bandelier, Adolf F.
1890–92 *Final Report of Investigations among the Indians of the Southwestern United States, Carried on Mainly in the Years 1880 to 1885.* 2 vols. Papers of the Archaeological Institute of America, American Series, nos. 3 and 4. Cambridge, Mass.

Barlow, R. H.
1949 *The Extent of the Empire of the Culhua Mexica.* Ibero-Americana, no. 28.

Beaglehole, Ernest
1937 *Notes on Hopi Economic Life.* Yale University Publications in Anthropology, no. 15. New Haven.

Beckerman, Stephen
1983 Does Swidden Ape the Jungle? *Human Ecology* 11:1–12.

Benavides, Antonio, and Linda Manzanilla
1985 Unidades Habitacionales Excavadas en Coba, Quintana Roo. In *Arquitectura y Arqueología: Metodologias en la Cronología de Yucatán,* edited by Dominique Michelet, pp. 69–76. Centre d'Etudes Mexicaines et Centramericaines, Mexico, D.F.

Bergman, Roland W.
1980 *Amazon Economics: The Simplicity of Shipibo Indian Wealth.* Dellplain Latin American Studies, no 6. Department of Geography, Syracuse University, New York. University Microfilms, Ann Arbor.

Binford, Lewis R.
1979 Organization and Formation Processes: Looking at Curated Technologies. *Journal of Anthropological Research* 35:255–73.

Black, Kevin D.
1979 Settlement Patterns in the Zapotitán Valley, El Salvador. Master's thesis, Department of Anthropology, University of Colorado, Boulder.

1983   The Zapotitán Valley Archaeological Survey. In *Archaeology and Volcanism in Central America: The Zapotitán Valley of El Salvador,* edited by Payson D. Sheets, pp. 62–97. University of Texas Press, Austin.

Blaikie, P., and Harold C. Brookfield
1986   *Land Degradation and Society.* Methuen, New York.

Bloom, Paul R., Mary Pohl, and Julie K. Stein
1985   Analysis of Sedimentation and Agriculture along the Rio Hondo, Northern Belize. In *Prehistoric Lowland Maya Environment and Subsistence Economy,* edited by Mary Pohl, pp. 21–34. Peabody Museum of Archaeology and Ethnology, Harvard University, Cambridge.

Bohannan, Paul
1954   *Tiv Farm and Settlement.* Colonial Research Studies, no. 15. Her Majesty's Stationery Office, London.

Bohrer, Vorsila L.
1960   Zuni Agriculture. *El Palacio* 67:181–202.
1970   Ethnobotanical Aspects of Snaketown: A Hohokam Village in Southern Arizona. *American Antiquity* 35:413–30.

Boserup, Ester
1965   *The Conditions of Agricultural Growth: The Economics of Agrarian Change under Population Pressure.* Aldine, Chicago.

Bradfield, Maitland
1971   *The Changing Pattern of Hopi Agriculture.* Occasional Paper of the Royal Anthropological Institute of Great Britain and Ireland no. 30. London.

Bray, Warwick
1972   Land-use, Settlement Pattern and Politics in Prehispanic Middle America: A Review. In *Man, Settlement and Urbanism,* edited by Peter J. Ucko, Ruth Tringham, and G. W. Dimbleby, pp. 909–26. Duckworth, London.

Broadbent, S. M.
1987   The Chibcha Raised-Field System in the Sabana de Bogota, Columbia: Further Investigations. In *Pre-Hispanic Agricultural Fields in the Andean Region,* edited by William M. Denevan, Kent Mathewson, and Gregory Knapp, pp. 425–42. BAR International Series 359. Oxford.

Brookfield, Harold C.
1962   Local Study and Comparative Method: An Example from Central New Guinea. *Annals of the Association of American Geographers* 52:242–52.

REFERENCES

   1972   Intensification and Disintensification in Pacific Agriculture: A Theoretical Approach. *Pacific Viewpoint* 13:30–84.

Brookfield, Harold C., and D. Hart
   1971   *Melanesia, a Geographical Interpretation of an Island World.* Methuen, London.

Bryan, Kirk
   1929   Flood Water Farming. *Geographical Review* 19:444–56.

Buchanan, K. M., and J. C. Pugh
   1955   *Land and People in Nigeria: The Human Geography of Nigeria and Its Environmental Background.* University of London Press, London.

Bugé, David E.
   n.d.a   Preliminary Report: 1978 Excavations at NM-01-1407, Ojo Caliente, New Mexico. Ms. on file, Department of Sociology/Anthropology, Occidental College, Los Angeles.
   n.d.b   Preliminary Report: 1978 Excavations at Ponsipa-akeri, Ojo Caliente, New Mexico. Ms. on file, Department of Sociology/Anthropology, Occidental College, Los Angeles.
   1981   Prehistoric Subsistence Strategies in the Chama Region, Northern New Mexico. Paper presented at the 46th Annual Meeting of the Society for American Archaeology, San Diego.
   1984   Prehistoric Subsistence Strategies in the Chama Region, Northern New Mexico. In *Prehistoric Agricultural Strategies in the Southwest*, edited by Suzanne K. Fish and Paul R. Fish, pp. 27–34. Anthropological Research Papers, no. 33. Arizona State University, Tempe.

Bullard, William R., Jr.
   1952   Boundary Walls and House Plots at Mayapan. Carnegie Institution of Washington Current Reports, vol. 1, no. 13. Washington, D.C.
   1965   *Stratigraphic Excavations at San Estevan, Northern British Honduras.* Occasional Paper no. 9. Royal Ontario Museum, University of Toronto.

Bye, Robert A., Jr.
   1979   Incipient Domestication of Mustards in Northwest Mexico. *Kiva* 44:237–56.

Calnek, Edward
   1973   The Localization of the Sixteenth Century Map Called the Maguey Plan. *American Antiquity* 38:190–95.
   1982   Patterns of Empire Formation in the Valley of Mexico, Late Postclassic Period, 1200–1521. In *The Inca and Aztec States 1400–1800*, edited by George Collier, Renato Rosaldo, and J. Wirth, pp. 43–62. Academic Press, New York.

Carneiro, Robert L.
1961 Slash-and-burn Cultivation among the Kuikuru and Its Implications for Cultural Development in the Amazon Basin. In *The Evolution of Horticultural Systems in Native South America: Causes and Consequences—A Symposium,* edited by Johannes Wilbert, pp. 47–67. Sociedad de Ciencias Naturales, Caracas.

Carrasco, Pedro
1976 The Joint Family in Ancient Mexico: The Case of Molotla. In *Essays on Mexican Kinship,* edited by H. C. Nutini, Pedro Carrasco, and James Taggart, pp. 45–64. University of Pittsburgh Press, Pittsburgh.

Carter, George F.
1945 *Plant Geography and Culture History in the American Southwest.* Viking Fund Publications in Anthropology, no. 5. New York.

Castetter, Edward F., and Willis H. Bell
1942 *Pima and Papago Indian Agriculture.* Interamericana Studies, no. 1. University of New Mexico Press, Albuquerque.

Charlton, Thomas H.
1970 Contemporary Settlement Patterns: The Cerro Gordo–North Slope and Upper Valley Areas. In *The Natural Environment, Contemporary Occupation, and 16th Century Population of the Valley. The Teotihuacan Valley Project Final Report,* vol. 1, edited by William T. Sanders, A. Kovar, Thomas H. Charlton, and Richard A. Diehl, pp. 181–236. Occasional Papers in Anthropology, no. 3. Department of Anthropology, Pennsylvania State University, University Park.

Chase, J. E.
1981 The Sky Is Falling: The San Martin Volcanic Eruption and Its Effects on the Olmec at Tres Zapotes, Veracruz. *Vinculos* 7:53–69.

Childe, V. Gordon
1937 *Man Makes Himself.* Watts, London.

Clarke, William C.
1971 *Place and People: An Ecology of a New Guinean Community.* University of California Press, Berkeley and Los Angeles.

Clary, Karen
1987 Analysis of Pollen Samples for Museum of New Mexico Project 41.374. Ms. on file, Laboratory of Anthropology, Museum of New Mexico, Santa Fe.

Cobean, Robert H., and A. G. Mastache
1989 The Late Classic and Early Postclassic Chronology of the Tula Region.

REFERENCES

    In *Tula of the Toltecs,* edited by Dan M. Healan, pp. 34–46. University of Iowa Press, Iowa City.

Cobean, Robert H., A. G. Mastache, A. Crespo, and C. Diaz
    1981   La Cronología de la región de Tula. In *Interaccion Cultural en Mexico Central,* compiled by Evelyn Rattray, Jaime Litvak, and C. Diaz, pp. 187–214. Universidad Nacional Autónoma de Mexico, Mexico City.

Codex Mendoza
    1980   [1541–1542] *Codice Mendoza.* Reprint of 1925 facsimile edition by F. del Paso y Troncoso, with notes by J. Galindo y Villa. Editorial Innovación, Mexico City.

Codice Xolotl
    1980   [16th century] *Codice Xolotl.* Edited by Charles E. Dibble. Universidad Nacional Autónoma de Mexico, Mexico City.

Conklin, Harold C.
    1957   *Hanunóo Agriculture: A Report of an Integral System of Shifting Cultivation in the Philippines.* Food and Agriculture Organization of the United Nations, Rome.

Cook, Scott
    1982   *Zapotec Stoneworkers: The Dynamics of Rural Simple Commodity Production in Modern Mexican Capitalism.* University Press of America, Washington, D.C.

Cordell, Linda S.
    1984   *Prehistory of the Southwest.* Academic Press, New York.

Cordell, Linda S., Amy C. Earls, and Martha R. Binford
    1984   Subsistence Systems in the Mountainous Settings of the Rio Grande Valley. In *Prehistoric Agricultural Strategies in the Southwest,* edited by Suzanne K. Fish and Paul R. Fish, pp. 233–41. Anthropological Research Papers, no. 33. Arizona State University, Tempe.

Cordell, Linda S., and Fred Plog
    1979   Escaping the Confines of Normative Thought: A Reevaluation of Puebloan Prehistory. *American Antiquity* 44:405–29.

Corey, A. T., and W. D. Kemper
    1968   *Conservation of Soil Water by Gravel Mulches.* Hydrology Papers, no. 30. Colorado State University, Fort Collins.

Covich, Alan P., and Norton H. Nickerson
    1966   Studies of Cultivated Plants in Choco Dwelling Clearings, Darien, Panama. *Economic Botany* 20:285–301.

Cowgill, Ursula
   1961   *Soil Fertility and Ancient Maya.* Transactions of the Connecticut Academy of Arts and Science, no. 42.
   1962   An Agricultural Study of the Southern Maya Lowlands. *American Anthropologist* 64:273–86.

Culbert, T. Patrick, Laura J. Levi, and Luis Cruz
   1989   The Rio Azul Agronomy Program, 1986 Season. In *Rio Azul Reports Number 4, the 1986 Season,* edited by R. E. W. Adams, pp. 189–214. University of Texas, San Antonio.

Cushing, Frank Hamilton
   1920   *Zuni Breadstuff.* Indian Notes and Monographs, no. 8. Museum of the American Indian, Heye Foundation, New York.

Daugherty, Howard Edward
   1969   *Man-Induced Ecological Changes in El Salvador.* Ph.D. diss., Department of Anthropology, University of California at Los Angeles. University Microfilms, Ann Arbor.

Davidson, Stanley, R. Passmore, J. F. Brock, and A. S. Truswell
   1979   *Human Nutrition and Dietetics.* Churchill Livingstone, Edinburgh.

Deal, Michael
   1983   Ceramic Systems of the Contemporary Tzeltal Maya: An Ethno-Archaeological Study. Ph.D. diss., Department of Archaeology, Simon Frazer University, Burnaby, British Columbia, Canada.

Dean, Jeffrey S., Robert C. Euler, George J. Gumerman, Fred Plog, Richard H. Hevley, and Thor N. V. Karlstrom
   1985   Human Behavior, Demography, and Paleoenvironment on the Colorado Plateaus. *American Antiquity* 50:537–54.

Denevan, William M.
   1970   Aboriginal Drained-Field Agriculture in the Americas. *Science* 169:647–54.
   1982   Hydraulic Agriculture in the American Tropics: Forms, Measures, and Recent Research. In *Maya Subsistence: Studies in Memory of Dennis E. Puleston,* edited by Kent V. Flannery, pp. 181–203. Academic Press, New York.

Denevan, William M., Kent Mathewson, and Gregory Knapp (editors)
   1987   *Pre-Hispanic Agricultural Fields in the Andean Region.* BAR International Series 359. Oxford.

Denevan, William M., and Karl Schwerin
   1978   Adaptive Strategies in Karinya Subsistence, Venezuelan Llanos. *Antropologica* 50:3–91.

REFERENCES

Denevan, William M., and B. L. Turner II
1974   Forms, Functions, and Associations of Raised Fields in the Old World Tropics. *Journal of Tropical Geography* 39:24–33.

de Schlippe, Pierre
1956   *Shifting Cultivation in Africa: The Zende System of Agriculture.* Routledge and Kegan Paul, London.

Diehl, Richard A.
1981   Tula. In *Supplement to the Handbook of Middle American Indians,* vol. 1, edited by Jeremy A. Sabloff, pp. 277–95. University of Texas Press, Austin.

1983   *Tula: The Toltec Capital of Ancient America.* Thames and Hudson, London.

Donkin, R. A.
1970   Pre-Columbian Field Implements and Their Distribution in the Highlands of Middle and South America. *Anthropos* 65:505–29.

1979   *Agricultural Terracing in the Aboriginal New World.* Viking Fund Publications in Anthropology, no. 56, New York.

Doolittle, William E.
1980   Aboriginal Agricultural Development in the Valley of Sonora, Mexico. *Geographical Review* 70:328–42.

1983   Agricultural Expansion in a Marginal Area of Mexico. *Geographical Review* 73:301–13.

1984a Agricultural Change as an Incremental Process. *Annals of the Association of American Geographers* 74:124–37.

1984b Cabeza de Vaca's Land of Maize: An Assessment of Its Agriculture. *Journal of Historical Geography* 10:246–62.

1984c Settlements and the Development of "Statelets" in Sonora, Mexico. *Journal of Field Archaeology* 11:13–24.

1988   *Pre-Hispanic Occupance in the Valley of Sonora, Mexico: Archaeological Confirmation of Early Spanish Reports.* Anthropological Papers of the University of Arizona, no. 48. Tucson.

Dougherty, Julia D.
1980   *An Archaeological Evaluation of Tsiping Ruin (AR 03-10-01-01).* Cultural Resources Report, no. 1. Santa Fe National Forest, Santa Fe.

Doyel, David E., and Fred Plog (editors)
1980   *Current Issues in Hohokam Prehistory: Proceedings of a Symposium.* Anthropological Research Papers, no. 23. Arizona State University, Tempe.

Dozier, Edward P.
1970 *The Pueblo Indians of North America*. Holt, Rinehart and Winston, New York.

Drennan, R. D.
1988 Household Location and Compact Versus Dispersed Settlement in Prehispanic Mesoamerica. In *Household and Community in the Mesoamerican Past*, edited by Richard R. Wilk and Wendy Ashmore, pp. 273–93. University of New Mexico Press, Albuquerque.

Earle, T. K.
1976 A Nearest-Neighbor Analysis of Two Formative Settlement Systems. In *The Early Mesoamerican Village*, edited by Kent V. Flannery, pp. 196–223. Academic Press, New York.

Earnest, Howard H., Jr.
1976 Investigaciones Efectuadas por el Proyecto No. 1, Programa de Rescate Arqueológico Cerrón Grande, en el Hacienda Santa Bárbara, Departamento de Chalatenango. *Publicaciones del Ministerio de Educación, Anales del Museo National "David J. Guzmán"* 49:64–66. San Salvador.

Eggan, Fred
1950 *Social Organization of the Western Pueblos*. Aldine, Chicago.

Eidt, Robert C.
1973 A Rapid Chemical Field Test for Archaeological Site Surveying. *American Antiquity* 38:206–10.

1977 Detection and Examination of Anthrosols by Phosphate Analysis. *Science* 197:1327–33.

1984 *Advances in Abandoned Settlement Analysis: Application to Prehistoric Anthrosols in Colombia, South America*. Center for Latin America, University of Wisconsin, Milwaukee.

Ellis, Florence H.
1967 Water Rights Studies of Nambe, Pojoaque, Tesuque, and San Ildefonso Pueblos. USDI Bureau of Indian Affairs, Washington, D.C. Ms. on file, Laboratory of Anthropology, Museum of New Mexico, Santa Fe.

1970 Irrigation and Water Works in the Rio Grande. Paper presented at the symposium, Water Control Systems, 1970 Pecos Conference, Santa Fe. Ms. on file, Laboratory of Anthropology, Museum of New Mexico, Santa Fe.

1975 Highways to the Past: The Valleys of the Rio Chama and Rio Gallina. *New Mexico Magazine* 53(5): 18–25, 38–40.

REFERENCES

1979 Summaries of the Histories of Water Use and the Tewa Culture of the Pojoaque Valley Pueblos. Ms. on file, Laboratory of Anthropology, Museum of New Mexico, Santa Fe.

Erickson, Clark L.
1984 Applications of Prehistoric Andean Technology: Experiments in Raised Field Agriculture, Huatta, Lake Titicaca: 1981–1982. In *Prehistoric Intensive Agriculture in the Tropics*, edited by Ian S. Farrington. BAR International Series 232. Oxford.

Evans, Susan T.
1980 A Settlement System Analysis of the Teotihuacan Region, Mexico, A.D. 1350–1520. Ph.D. diss., Department of Anthropology, Pennsylvania State University, University Park. University Microfilms, Ann Arbor.

1985 The Cerro Gordo Site: A Rural Settlement of the Aztec Period in the Basin of Mexico. *Journal of Field Archaeology* 12:1–18.

1986 Analysis of the Surface Sample Ceramics. In *The Toltec Period Occupation of the Valley. Part 1, Excavations and Ceramics. The Teotihuacan Valley Project Final Report*, vol. 4, edited by William T. Sanders, A. Kovar, Thomas H. Charlton, and Richard A. Diehl, pp. 283–366. Occasional Papers in Anthropology, no. 3. Department of Anthropology, Pennsylvania State University, University Park.

1988 Cihuatecpan: The Village in Its Ecological and Historical Context. In *Excavations at Cihuatecpan, An Aztec Village in the Teotihuacan Valley*, edited by Susan T. Evans, pp. 1–49. Vanderbilt University Publications in Anthropology, no. 36. Nashville.

1989 House and Household in the Aztec World: The Village of Cihuatecpan. In *Households and Communities*, edited by S. MacEachern, D. Archer, and R. Garvin, pp. 430–40. Proceedings of the 21st Annual Chacmool Conference. Archaeological Association of the University of Calgary, Calgary.

1990 Household Division of Labor at the Aztec Period Village of Cihuatecpan (Mexico). Paper presented at the symposium, Household Organization: Empirical Approaches to Continuity and Change, Annual Meeting of the American Anthropological Association, New Orleans.

n.d. Aztec Household Organization and Village Administration. In *Household, Compound, and Residence: Studies of Prehispanic Domestic Units in Western Mesoamerica*, edited by K. G. Hirth and Robert S. Santley. Telford Press, West Caldwell, New Jersey.

Evans, Susan T., and Elliot M. Abrams
1988 Archaeology at the Aztec Period Village of Cihuatecpan, Mexico. In *Excavations at Cihuatecpan, An Aztec Village in the Teotihuacan Valley*,

edited by Susan T. Evans, pp. 50–233. Vanderbilt University Publications in Anthropology, no. 36. Nashville.

Evans, Susan T., and A. C. Freter
  1989   Hydration Analysis of Obsidian from Cihuatecpan, an Aztec Period Village in Mexico. Paper presented at the 54th Annual Meeting of the Society for American Archaeology, Atlanta.

Fairbourn, Merle L.
  1973   Effect of Gravel Mulch on Crop Yields. *Colorado Rancher and Farmer* 24(3): 925–28.

Fallon, Denise P., and Karen Wening
  1987   *Howiri: Excavation at a Northern Rio Grande Biscuit Ware Site.* Laboratory of Anthropology Notes, no. 216b. Museum of New Mexico, Santa Fe.

Farrington, Ian S.
  1985a   The Wet, the Dry and the Steep. Archaeological Imperatives and the Study of Agricultural Intensification. In *Prehistoric Intensive Agriculture in the Tropics, Part I*, edited by Ian S. Farrington, pp. 1–10. BAR International Series 232. Oxford.

Farrington, Ian S. (editor)
  1985b   *Prehistoric Intensive Agriculture in the Tropics, Parts I and II.* BAR International Series 232. Oxford.

Farrington, Ian S., and James Urry
  1985   Food and the Early History of Cultivation. *Journal of Ethnobiology* 5:143–57.

Farriss, Nancy M.
  1984   *Maya Society Under Colonial Rule: The Collective Enterprise of Survival.* Princeton University Press, Princeton.

Fiero, Kathleen
  1978   *Prehistoric Garden Plots Along the Lower Rio Chama Valley: Archaeological Investigations at Sites LA 11830, LA 11831, and LA 11832, Rio Arriba County, New Mexico* (review draft). Laboratory of Anthropology Notes, no. 111e. Museum of New Mexico, Santa Fe.

Fish, Suzanne K., and Paul R. Fish (editors)
  1984   *Prehistoric Agricultural Strategies in the Southwest.* Anthropological Research Papers, no. 33. Arizona State University, Tempe.

Fish, Suzanne K., Paul R. Fish, and Christian Downum
  1983   Hohokam Terraces and Agricultural Production in the Tucson Basin. In *Prehistoric Agricultural Strategies in the Southwest*, edited by Su-

zanne K. Fish and Paul R. Fish, pp. 55–71. Anthropological Research Papers, no. 33. Arizona State University, Tempe.

Fittkau, E. J., and H. Klinge
1973 On Biomass and Trophic Structure of the Central Amazonian Rain Forest Ecosystem. *Biotropica* 5(1): 1–14.

Flannery, Kent V. (editor)
1976 *The Early Mesoamerican Village*. Academic Press, New York.
1982 *Maya Subsistence: Studies in Memory of Dennis E. Puleston*. Academic Press, New York.

Folan, William J., E. R. Kintz, and L. A. Fletcher
1983 *Coba: A Classic Maya Metropolis*. Academic Press, New York.

Follansbee, Robert, and H. J. Dean
1915 *Water Sources of the Rio Grande Basin 1888–1913*. U.S. Geological Survey, Water-Supply Paper no. 358. U.S. Government Printing Office, Washington, D.C.

Ford, Richard I.
1972 An Ecological Perspective on the Eastern Pueblos. In *New Perspectives on the Pueblos*, edited by Alfonso Ortiz, pp. 1–17. University of New Mexico Press, Albuquerque.
1981 Gardening and Farming Before A.D. 1000: Patterns of Prehistoric Cultivation North of Mexico. *Journal of Ethnobiology* 1:6–27.

Forde, C. Daryll
1931 Hopi Agriculture and Land Ownership. *Journal of the Royal Anthropological Institute of Great Britain and Ireland* 61:357–405.

Fowler, Melvin L.
1969 Middle Mississippian Agricultural Fields. *American Antiquity* 34:365–75.
1971 The Origin of Plant Cultivation in the Central Mississippi Valley: A Hypothesis. In *Prehistoric Agriculture*, edited by Stuart Struever, pp. 122–28. Natural History Press, Garden City, New York.

Fox, Richard G.
1977 *Urban Anthropology: Cities in Their Cultural Settings*. Prentice-Hall, Englewood Cliffs, New Jersey.

Freidel, David A.
1976 Late Postclassic Settlement Patterns on Cozumel Island, Quintana Roo. Ph.D. diss., Department of Anthropology, Harvard University, Cambridge.

REFERENCES

Freidel, David A., and Jeremy A. Sabloff
 1984 *Cozumel: Late Maya Settlement Patterns.* Academic Press, New York.

Freidel, David A., and Vernon Scarborough
 1982 Subsistence, Trade and Development of the Coastal Maya. In *Maya Subsistence: Studies in Memory of Dennis E. Puleston,* edited by Kent V. Flannery, pp. 131–55. Academic Press, New York.

Fry, Robert E.
 1983 The Ceramics of the Pulltrouser Area: Settlements and Fields. In *Pulltrouser Swamp: Ancient Maya Habitat, Agriculture, and Settlement in Northern Belize,* edited by B. L. Turner II and Peter D. Harrison, pp. 194–211. University of Texas Press, Austin.

Gauthier, Rory P., Patricia A. Prince, and Frances Joan Mathien
 1978 *An Archeological Sample Survey of Proposed Timber Sale Areas on the Picuris Pueblo Reservation.* Office of Contract Archeology, University of New Mexico, Albuquerque. Submitted to USDI Bureau of Indian Affairs, Albuquerque Area Office, Albuquerque.

Geertz, Clifford
 1963 *Agricultural Involution: The Process of Ecological Change in Indonesia.* Association of Asian Studies, University of California Press, Berkeley and Los Angeles.

Gibson, C.
 1964 *The Aztecs Under Spanish Rule.* Stanford University Press, Stanford.

Glassow, Michael A.
 1977 Population Aggregation and Systemic Change: Examples from the American Southwest. In *Explanation of Prehistoric Change,* edited by James N. Hill, pp. 185–214. University of New Mexico Press, Albuquerque.

 1980 *Prehistoric Agricultural Development in the Northern Southwest.* Anthropological Papers, no. 16. Ballena Press, Socorro, New Mexico.

 1984 Explaining Variations in Agricultural Settlement Systems in Northeastern New Mexico. In *Prehistoric Agricultural Strategies in the Southwest,* edited by Suzanne K. Fish and Paul R. Fish, pp. 199–214. Anthropological Research Papers, no. 33. Arizona State University, Tempe.

Gliessman, Stephen R., B. L. Turner II, F. J. Rosado May, and M. F. Amador
 1985 Ancient Raised Field Agriculture in the Maya Lowlands of Southeastern Mexico. In *Prehistoric Intensive Agriculture in the Tropics,* edited by Ian S. Farrington, pp. 97–112. BAR International Series 232. Oxford.

REFERENCES

Golomb, Berl, and Herbert M. Eder
1964 Landforms Made by Man. *Landscape* 14:47.

Golson, Jack
1977 Simple Tools and Complex Technology: Agriculture and Agricultural Implements in the New Guinea Highlands. In *Stone Tools as Cultural Markers: Change, Evolution and Complexity,* edited by R. V. S. Wright, pp. 154–61. Australian Institute of Aboriginal Studies, Canberra.

Golson, Jack, and Axel Steensberg
1985 The Tools of Agricultural Intensification in the New Guinea Highlands. In *Prehistoric Intensive Agriculture in the Tropics,* Part I, edited by Ian S. Farrington, pp. 347–84. BAR International Series 232. Oxford.

Gomez-Pompa, Arturo
1973 Ecology of the Vegetation of Veracruz. In *Vegetation and Vegetational History of Northern Latin America,* edited by A. Graham, pp. 73–148. Elsevier Scientific, Amsterdam.

Graham, Martha
1986 Tarahumara Mobility and Residence: An Ethnoarchaeological Study of Settlement. Proposal to the National Science Foundation. Department of Anthropology, University of New Mexico, Albuquerque.

Greenlee, Robert
n.d. Archaeological Sites in the Chama Valley, and Report on Excavations at Tsama, 1929–1933. Ms. on file, Laboratory of Anthropology, Museum of New Mexico, Santa Fe.

Gregor, Thomas
1977 *Mehinaku: The Drama of Daily Life in a Brazilian Indian Village.* University of Chicago Press, Chicago.

Grigg, D. B.
1974 *The Agricultural Systems of the World: An Evolutionary Approach.* Cambridge University Press, London.

Guaman Poma de Ayala, Felipe
1936 [ca. 1613] *Nueva Corónica y Buen Gobierno: Codex Péruvien Illustré.* Institut d'ethnologie, Paris. [Facsimile edited by P. Rivet.]

Hack, John T.
1942 *The Changing Physical Environment of the Hopi Indians of Arizona.* Reports of the Awatovi Expedition 1, Papers of the Peabody Museum of Archaeology and Ethnology, vol. 35, no. 1. Harvard University, Cambridge.

Hackenberg, Robert A.
　1962　Economic Alternatives in Arid Lands: A Case Study of the Pima and Papago Indians. *Ethnology* 1:186–96.

Hammond, George P., and Agapito Rey
　1966　*The Rediscovery of New Mexico: The Explorations of Chamuscado, Espejo, Castaño de Sosa, Morlete, and Leyva de Bonilla and Humana*. University of New Mexico Press, Albuquerque.

Hammond, Norman
　1974　Preclassic to Postclassic in Northern Belize. *Antiquity* 48:177–89.
　1975　Maya Settlement Hierarchy in Northern Belize. *Contributions of the University of California Archaeological Research Facility* 27:40–55.
　1978　The Myth of the Milpa: Agricultural Expansion in the Maya Lowlands. In *Pre-Hispanic Maya Agriculture*, edited by Peter D. Harrison and B. L. Turner II, pp. 23–34. University of New Mexico Press, Albuquerque.
　1985　*Nohmul: A Prehistoric Maya Community in Belize, Excavations 1973–1983*. BAR International Series 250. Oxford.

Hammond, Norman (editor)
　1973　*British Museum–Cambridge Corozal Project, 1973 Interim Report*. Centre of Latin American Studies, Cambridge.

Hammond, Norman, Sara Donaghey, Colleen Gleason, J. C. Staneko, Dirk Van Tuerenhout, and Laura J. Kosakowsky
　1987　Excavations at Nohmul, Belize, 1985. *Journal of Field Archaeology* 14:257–81.

Harner, Michael
　1972　*The Jivaro: People of the Sacred Waterfalls*. Doubleday and Natural History Press, Garden City, New York.

Harris, David
　1972　The Origins of Agriculture in the Tropics. *American Scientist* 60:180–93.

Harris, Marvin
　1971　*Culture, Man, and Nature: An Introduction to General Anthropology*. Thomas Y. Crowell, New York.

Harrison, Peter D.
　1977　The Rise of the Bajos and the Fall of the Maya. In *Social Process in Maya Prehistory: Essays in Memory of Sir Eric Thompson*, edited by Norman Hammond, pp. 469–508. Academic Press, London.
　1978　Bajos Revisited: Visual Evidence for One System of Agriculture. In *Pre-Hispanic Maya Agriculture*, edited by Peter D. Harrison and B. L. Turner II, pp. 247–53. University of New Mexico Press, Albuquerque.

REFERENCES

1983   The Pulltrouser Settlement Survey and Mapping of Kokeal. In *Pulltrouser Swamp: Ancient Maya Habitat, Agriculture and Settlement in Northern Belize*, edited by B. L. Turner II and Peter D. Harrison, pp. 140–57. University of Texas Press, Austin.

1988   Functional Influences on Settlement Pattern in the Communities of Pulltrouser Swamp, Northern Belize. Paper presented at the 21st Annual Chacmool Conference, Calgary, Canada.

1990   The Revolution in Ancient Maya Subsistence. In *Vision and Revision in Maya Studies*, edited by Flora S. Clancy and Peter D. Harrison, pp. 99–113. University of New Mexico Press, Albuquerque.

Harrison, Peter D., and B. L. Turner II (editors)
1978   *Pre-Hispanic Maya Agriculture*. University of New Mexico Press, Albuquerque.

Hart, William J. E., Jr.
1983   Classic to Postclassic Tephra Layers Exposed in Archaeological Sites, Eastern Zapotitán Valley. In *Archaeology and Volcanism in Central America: The Zapotitán Valley of El Salvador*, edited by Payson D. Sheets, pp. 44–51. University of Texas Press, Austin.

Hart, William J. E., Jr., and Virginia Steen-McIntyre
1983   Tierra Blanca Joven Tephra from the AD 260 Eruption of Ilopango Caldera. In *Archaeology and Volcanism in Central America: The Zapotitán Valley of El Salvador*, edited by Payson D. Sheets, pp. 14–34. University of Texas Press, Austin.

Harvey, H.
1985   Household and Family Structure in Early Colonial Tepelaoztoc. *Estudios de Cultura Nahuatl* 18:275–94.

Haury, Emil W.
1937   The Snaketown Canal. In *Excavations at Snaketown, Vol. 1: Material Culture*, by Harold S. Gladwin, Emil W. Haury, Edwin B. Sayles, and Nora Gladwin, pp. 50–58. Medallion Papers, no. 25. Gila Pueblo, Globe, Arizona.

1976   *The Hohokam, Desert Farmers and Craftsmen: Excavations at Snaketown, 1964–1965*. University of Arizona Press, Tucson.

Haviland, William A.
1981   Dower Houses and Minor Centers at Tikal, Guatemala: An Investigation into the Identification of Valid Units in Settlement Hierarchies. In *Lowland Maya Settlement Patterns*, edited by Wendy Ashmore, pp. 89–117. School of American Research and University of New Mexico Press, Albuquerque.

1985   *Excavations in Small Residential Groups of Tikal: Groups 4F-1 and 4F-2.* Tikal Reports, no. 19. University Museum, University of Pennsylvania, Philadelphia.

Hawkes, J. G.
1969   The Ecological Background of Plant Domestication. In *The Domestication of Plants and Animals,* edited by Peter J. Ucko and G. W. Dimbleby, pp. 17–29. Aldine-Atherton, Chicago.

Hayden, Brian, and Aubrey Cannon
1982   The Corporate Group as an Archaeological Unit. *Journal of Anthropological Archaeology* 1:132–58.
1983   Where the Garbage Goes: Refuse Disposal in the Maya Highlands. *Journal of Anthropological Archaeology* 2:117–63.
1984   *The Structure of Material Systems: Ethnoarchaeology in the Maya Highlands.* SAA Papers, no. 3. Society for American Archaeology, Washington, D.C.

Hayes, Alden C.
1981   A Survey of Chaco Canyon Archaeology. In *Archaeological Surveys of Chaco Canyon,* by Alden C. Hayes, David M. Brugge, and W. James Judge, pp. 1–68. Chaco Canyon Studies Publications in Anthropology, no. 18a., USDI National Park Service, U.S. Government Printing Office, Washington, D.C.

Healan, Dan M., Robert H. Cobean, and Richard A. Diehl
1989   Synthesis and Conclusions. In *Tula of the Toltecs,* edited by Dan M. Healan, pp. 239–51. University of Iowa Press, Iowa City.

Healy, Paul F., J. D. H. Lambert, J. T. Arnason, and R. J. Hebda
1983   Caracol, Belize: Evidence of Ancient Maya Agricultural Terraces. *Journal of Field Archaeology* 10:397–410.

Hellige-Truog
n.d.   *Directions* (no. 697-18), Hellige-Truog Combination Soil Tester. Hellige, Garden City, New York.

Hellmuth, Nicholas M.
1977   Cholti-Lacandon and Peten-Itza Agriculture, Settlement Pattern, and Population. In *Social Process in Maya Prehistory,* edited by Norman Hammond, pp. 421–28. Academic Press, New York.

Herrera y Tordesillas, Antonio de
1726–27 *Historia general de los hechos de los Indias Ocidentales.* Imprenta Real de Nicolas Rodriquez, Madrid.

REFERENCES

Hester, Joseph A.
1954 Natural and Cultural Bases of Ancient Maya Subsistence Economy. Ph.D. diss., Department of Anthropology, University of California, Los Angeles.

Hewett, Edgar Lee
1906 *Antiquities of the Jemez Plateau, New Mexico.* Bureau of American Ethnology Bulletin no. 32. Smithsonian Institution, Washington, D.C.
1953 *Pajarito Plateau and Its Ancient People.* Revised by Bertha P. Dutton. School of American Research and University of New Mexico Press, Albuquerque.

Hibben, Frank C.
1937 *Excavation of the Riana Ruin and Chama Valley Survey.* University of New Mexico Bulletin no. 300, Anthropological Series 2(1). Albuquerque.

Hill, W. W.
1982 *An Ethnography of Santa Clara Pueblo, New Mexico.* University of New Mexico Press, Albuquerque.

Hiraoka, Mario
1986 Zonation of Mestizo Riverine Farming Systems in Northeast Peru. *National Geographic Research* 2:354–71.

Hodge, Frederick W.
1893 Prehistoric Irrigation in Arizona. *American Anthropologist* 6:323–30.

Hubbard, B.
1878 Ancient Garden Beds in Michigan. *American Antiquarian* 1:1–9.

Hull, Kathleen L.
1987 Identification of Cultural Site Formation Processes through Microdebitage Analysis. *American Antiquity* 52:772–83.

Hummer, Anne G.
1983 Ground Stone of the Zapotitán Valley. In *Archaeology and Volcanism in Central America: The Zapotitán Valley of El Salvador,* edited by Payson D. Sheets, pp. 229–51. University of Texas Press, Austin.

Ixtlilxochitl, F. de A.
1985 [1600–1640] *Obras Historicas,* vols. 1 and 2, edited by E. O'Gorman. Universidad Nacional Autónoma de Mexico, Mexico City.

Jeançon, J. A.
1912 Ruins at Pesedeuinge. *Records of the Past* 11(1): 28–37.

  1923 *Excavations in the Chama Valley, New Mexico.* Bureau of American Ethnology Bulletin no. 81. Smithsonian Institution, Washington, D.C.

Johnson, Allen
  1983 Machiguenga Gardens. In *Adaptive Responses of Native Amazonians*, edited by Raymond B. Hames and William T. Vickers, pp. 29–64. Academic Press, New York.

Keeley, Lawrence H.
  1982 Hafting and Retooling: Effects on the Archaeological Record. *American Antiquity* 47:798–809.

Kelley, J. C.
  1971 Archaeology of the Northern Frontier. In *The Handbook of Middle American Indians*, vol. 2, edited by Robert Wauchope, pp. 768–801. University of Texas Press, Austin.

Kelly, Isabel, and Angel Palerm
  1952 *The Tajin Totonac.* U.S. Government Printing Office, Washington, D.C.

Kelsay, Richalene G.
  1983 *Quail Gardens Tract, La Costa, California: Soil Phosphate Report.* Report on file, Paul Chase and Associates, Escondido, California.
  1985 A Late Classic Lithic Finishing Station at Buenavista, Belize. Paper presented at the 50th Annual Meeting of the Society for American Archaeology, Denver.

Kidder, Alfred V.
  1924 *An Introduction to the Study of Southwestern Archaeology, with a Preliminary Account of the Excavations at Pecos.* Papers of the Southwestern Expedition, no. 1. Published for Phillips Academy by Yale University Press, New Haven.

Killion, Thomas W.
  1985 Horticultural Practices and Houselot Organization in the Sierra de los Tuxtlas: Building a Foundation for Archaeological Inference. Paper presented at the 50th Annual Meeting of the Society for American Archaeology, Denver.
  1987a *Agriculture and Residential Site Structure Among Campesinos in Southern Veracruz, Mexico: Building a Foundation for Archaeological Inference.* Ph.D. diss., Department of Anthropology, University of New Mexico, Albuquerque. University Microfilms, Ann Arbor.
  1987b The Use of Space around the Residence by Ancient Farmers on the Gulf Coast: Recent Research from the Site of Matacapan, Veracruz,

REFERENCES

        Mexico. Paper presented at the 52d Annual Meeting of the Society for American Archaeology, Toronto.

1990   Cultivation Intensity and Residential Site Structure: An Ethnographic Examination of Peasant Agriculture in the Sierra de los Tuxtlas, Veracruz, Mexico. *Latin American Antiquity* 1:191–215.

Killion, Thomas W., Jeremy A. Sabloff, Gair Tourtellot, and Nicholas Dunning
1989   Intensive Surface Collection of Residential Clusters at Terminal Classic Sayil, Yucatan, Mexico. *Journal of Field Archaeology* 16:273–94.

Kimber, Clarissa T.
1966   Dooryard Gardens of Martinique. *Yearbook of the Association of Pacific Coast Geographers* 28:97–118.
1973   Spatial Patterning in the Dooryard Gardens of Puerto Rico. *Geographical Review* 63:6–26.

Kolata, A. L.
1986   The Agricultural Foundations of the Tiwanaku State: A View from the Hinterland. *American Antiquity* 51:748–62.

Koldehoff, Brad
1987   The Cahokia Flake Tool Industry: Socioeconomic Implications for Late Prehistory in the Central Mississippi Valley. In *The Organization of Core Technology*, edited by J. K. Johnson and C. A. Morrow, pp. 151–86. Westview Press, Boulder.

Lamb, J., Jr., and J. E. Chapman
1943   Effect of Surface Stones on Erosion, Evaporation, Soil Temperature and Soil Moisture. *Journal of the American Society of Agronomy* 35:567–78.

Lambert, J. D. H., and J. T. Arnason
1983   Ancient Maya Land-Use and Potential Agricultural Productivity at Lamanai, Belize. In *Drained Field Agriculture in Central and South America*, edited by J. P. Darch, pp. 111–22. BAR International Series 189. Oxford.

Lang, Richard W.
1979   *An Archaeological Survey near the Confluence of the Chama and Ojo Caliente Rivers, Rio Arriba County, New Mexico.* Contract Archaeology Division Report no. 5. School of American Research, Santa Fe.
1980   *Archaeological Investigations at a Pueblo Agricultural Site, and Archaic and Puebloan Encampments on the Rio Ojo Caliente, Rio Arriba County, New Mexico.* Contract Archeology Division Report no. 7. School of American Research, Santa Fe.

1981 *A Prehistoric Pueblo Garden Plot on the Rio Ojo Caliente, Rio Arriba County, New Mexico: Ojo Caliente Site 7, Features 1–2*. Contract Archeology Division Report no. 65. School of American Research, Santa Fe.

Lange, Charles H.
1959 *Cochiti: A New Mexico Pueblo, Past and Present*. University of Texas Press, Austin.

Lathrap, Donald W.
1977 Our Father the Cayman, Our Mother the Gourd: Spinden Revisited or a Unitary Model for the Emergence of Agriculture in the New World. In *The Origins of Agriculture*, edited by C. A. Reed, pp. 713–51. Mouton, The Hague.

Lightfoot, Kent G., and Fred Plog
1984 Intensification Along the North Side of the Mogollon Rim. In *Prehistoric Agricultural Strategies in the Southwest*, edited by Suzanne K. Fish and Paul R. Fish, pp. 179–95. Anthropological Research Papers, no. 33. Arizona State University, Tempe.

Linares, Olga F.
1976 "Garden Hunting" in the American Tropics. *Human Ecology* 4:331–49.

Linares, Olga F., Payson D. Sheets, and E. Jane Rosenthal
1975 Prehistoric Agriculture in Tropical Highlands. *Science* 187:137–45.

Luebben, Ralph A.
1953 Leaf Water Site. In *Salvage Archaeology in the Chama Valley, New Mexico*, assembled by Fred Wendorf, pp. 9–33. Monographs of the School of American Research, no. 17. Santa Fe.

Lundell, Cyrus L.
1933 The Agriculture of the Maya. *Southwest Review* 19:65–77.

McAnany, Patricia A.
1986 *Lithic Technology and Exchange Among Wetland Farmers of the Eastern Maya Lowlands*. Ph.D. diss., Department of Anthropology, University of New Mexico, Albuquerque. University Microfilms, Ann Arbor.

1988 Effect of Lithic Procurement Strategies on Tool Curation and Recycling. *Lithic Technology* 17:3–11.

1989 Stone Tool Production and Exchange in the Eastern Maya Lowlands: The Consumer Perspective from Pulltrouser Swamp, Belize. *American Antiquity* 54:332–46.

n.d. K'axob: A Formative and Classic Period Settlement at Pulltrouser Swamp, Belize. Ms. in possession of author.

REFERENCES

McAnany, Patricia A., and Barry L. Isaac (editors)
    1989   *Prehistoric Maya Economies of Belize*. Research in Economic Anthropology, Supplement no. 4. JAI Press, Greenwich.

McCourt, D.
    1955   Infield and Outfield in Ireland. *Economic History Review* 7:369–76.

McQuarie, Harriet
    n.d.   Buried Structures at Pulltrouser Swamp. Ms. in possession of author.

Marshall, Michael P.
    1982   *Excavations at Nuestra Señora de Delores Pueblo (LA 677), a Prehistoric Settlement in the Tiguex Province*. Office of Contract Archeology, University of New Mexico, Albuquerque.

Masse, W. Bruce
    1979   An Intensive Survey of Prehistoric Dry Farming Systems Near Tumamoc Hill in Tucson, Arizona. *Kiva* 45:141–86.
    1981   Prehistoric Irrigation Systems in the Salt River Valley, Arizona. *Science* 214:408–15.

Matheny, Raymond
    1978   Northern Maya Lowland Water-Control Systems. In *Pre-Hispanic Maya Agriculture*, edited by Peter D. Harrison and B. L. Turner II, pp. 185–210. University of New Mexico Press, Albuquerque.

Meggers, Betty J.
    1971   *Amazonia: Man and Culture in a Counterfeit Paradise*. Aldine, Chicago.

Mera, H. P.
    1934   *A Survey of the Biscuit Ware Area in Northern New Mexico*. Technical Series Bulletin no. 86. Laboratory of Anthropology, Museum of New Mexico, Santa Fe.

Miksicek, Charles H.
    1983   Macrofloral Remains of the Pulltrouser Area: Settlements and Fields. In *Pulltrouser Swamp: Ancient Maya Habitat, Agriculture, and Settlement in Northern Belize*, edited by B. L. Turner II and Peter D. Harrison, pp. 94–104. University of Texas Press, Austin.

Moore, James L.
    1981   Prehistoric Soil and Water Conservation in the Middle Rio Puerco Valley. Master's thesis, Department of Anthropology, University of New Mexico, Albuquerque.

Moran, Emilio F.
    1979   *Human Adaptability: An Introduction to Ecological Anthropology*. Westview Press, Boulder.

Morley, Sylvanius G.
1946   *The Ancient Maya*. Stanford University Press, Stanford.

Morley, Sylvanius G., and G. W. Brainerd
1956   *The Ancient Maya*. Rev. ed. Stanford University Press, Stanford.

Motolinia (Fray T. de Benavente)
1971   [1536–43] *Memoriales*. Universidad Nacional Autónoma de Mexico, Mexico City.

Munsell Color
1976   *Munsell Book of Color*. Munsell Color, Kollmorgen Corporation, Baltimore.

Murphy, Sean
1989   Casting Organic Materials. In *1989 Archaeological Investigations at the Cerén Site, El Salvador: A Preliminary Report*, edited by Payson D. Sheets and Brian R. McKee, pp. 27–28. Department of Anthropology, University of Colorado, Boulder.

Nabhan, Gary Paul
1985   Native Crop Diversity in Aridoamerica: Conservation of Regional Gene Pools. *Economic Botany* 39:387–99.

Naroll, Raoul
1962   Floor Area and Settlement Pattern. *American Antiquity* 27:490–502.

Nelson, M. C.
1987   Site Content and Structure: Quarries and Workshops in the Maya Highlands. In *Lithic Studies Among the Contemporary Highland Maya*, edited by Brian Hayden, pp. 120–47. University of Arizona Press, Tucson.

Nelson, Nels C.
1914   *Pueblo Ruins of the Galisteo Basin, New Mexico*. Anthropological Papers of the American Museum of Natural History, vol. 15, no. 1. New York.

Netting, Robert M.
1968   *Hill Farmers of Nigeria: Cultural Ecology of the Kofyar of the Jos Plateau*. University of Washington Press, Seattle.

1977   Maya Subsistence: Mythologies, Analogies, Possibilities. In *The Origins of Maya Civilization*, edited by R. E. W. Adams, pp. 299–333. University of New Mexico Press, Albuquerque.

Nicholas, Linda M.
1981   Irrigation and Sociopolitical Development in the Salt River Valley, Arizona: An Examination of Three Prehistoric Canal Systems. Mas-

ter's thesis, Department of Anthropology, Arizona State University, Tempe.

Nichols, Deborah
1987 Risk and Agricultural Intensification During the Formative Period in the Northern Basin of Mexico. *American Anthropologist* 89:596–616.

Niñez, Vera K.
1986 *Food Production for Home Consumption: Nature and Function of Gardens in Household Economies.* International Potato Center, Lima.

Nye, P. H., and D. J. Greenland
1960 The Soil under Shifting Cultivation. *Technical Communication* no. 51. Commonwealth Bureau of Soils. Harpenden, Commonwealth Agricultural Bureau.

Olson, Gerald W.
1983a Appendix 4-A: Soil Test Laboratory Data. In *Archaeology and Volcanism in Central America: The Zapotitán Valley of El Salvador,* edited by Payson D. Sheets, pp. 57–61. University of Texas Press, Austin.
1983b An Evaluation of Soil Properties and Potentials in Different Volcanic Deposits. In *Archaeology and Volcanism in Central America: The Zapotitán Valley of El Salvador,* edited by Payson D. Sheets, pp. 52–56. University of Texas Press, Austin.

Oviedo y Valdes, G. H.
1853 *História General y Nacional de los Indes.* Vol. 3. Memorial de la Academia de la História, Madrid.

Page, Gordon B.
1940 Hopi Agricultural Notes. Ms. on file, USDA Soil Conservation Service, Washington, D.C.

Palerm, Angel
1952 *La Civilización Urbana.* História Mexicana 2. Mexico, D.F.
1955 The Agricultural Basis of Urban Civilization in Mesoamerica. In *Irrigation Civilizations: A Comparative Study,* edited by J. H. Steward, pp. 28–42. Social Science Monographs, no. 1, Pan American Union, Washington, D.C.

Palerm, Angel, and Eric R. Wolf
1961 La Agricultura y el Desarrollo de la Civilizacion en Mesoamerica. *Revista Interamericana de Ciencias Sociales,* Segunda epoca, vol. 1, Pan American Union, Washington, D.C.

Parsons, James J., and William M. Denevan
1967 Pre-Columbian Ridged Fields. *Scientific American* 217:92–100.

REFERENCES

Parsons, Jeffrey R.
1971 *Prehistoric Settlement Patterns in the Texcoco Region, Mexico*. Memoirs of the Museum of Anthropology, no. 3, University of Michigan, Ann Arbor.

1976 The Role of Chinampa Agriculture in the Food Supply of Aztec Tenochtitlan. In *Cultural Change and Continuity*, edited by C. Cleland, pp. 233–57. Academic Press, New York.

Parsons, Jeffrey R., and Mary H. Parsons
1985 *Otomi Maguey Utilization: An Ethnoarchaeological Perspective*. Preliminary report to the National Geographic Society. University of Michigan Museum of Anthropology, Ann Arbor.

1987 *Maguey Sap Utilization in Highland Central Mexico: An Archaeological Ethnography*. Preliminary report to the National Geographic Society. University of Michigan Museum of Anthropology, Ann Arbor.

Parsons, Mary H.
1972 Spindle Whorls from the Teotihuacan Valley, Mexico. In *Miscellaneous Studies in Mexican Prehistory*, edited by M. W. Spence, Jeffrey R. Parsons, and Mary H. Parsons, pp. 81–117. Anthropological Papers of the Museum of Anthropology, no. 45, University of Michigan Museum of Anthropology, Ann Arbor.

Peckham, Stewart
1981 The Palisade Ruin. In *Collected Papers in Honor of Erik Kellerman Reed*, edited by A. H. Schroeder, pp. 113–47. Papers of the Archaeological Society of New Mexico, no. 6.

Pennington, Campbell W.
1963 *The Tarahumar of Mexico: Their Environment and Material Culture*. University of Utah Press, Salt Lake City.

1969 *The Tepehuan of Chihuahua: Their Material Culture*. University of Utah Press, Salt Lake City.

1980 *The Pima Bajo of Central Sonora, Mexico, Vol. 1: The Material Culture*. University of Utah Press, Salt Lake City.

Pfefferkorn, Ignaz
1989 *Sonora: A Description of the Province*. Translated and edited by Theodore E. Treutlein. University of Arizona Press, Tucson.

Plog, Fred, and C. Garrett
1972 Explaining Variability in Southwestern Water Control Systems. In *Contemporary Archaeology*, edited by Mark P. Leone, pp. 280–88. Southern Illinois University Press, Carbondale.

REFERENCES

Pool, Christopher A., Ponciano Ortiz, Z. Salazar, J. Martinez, and Robert S. Santley
   1986   *Final Field Report, Matacapan Project: 1986 Season.* Final report to the National Science Foundation and the Instituto Nacional de Antropologia e História.

Pope, Kevin O., and Bruce H. Dahlin
   1989   Ancient Maya Wetland Agriculture: New Insights from Ecological and Remote Sensing Research. *Journal of Field Archaeology* 16:87–106.

Pospisil, Leopold
   1963   *Kapauku Papuan Economy.* Yale University Publications in Anthropology, no. 67. New Haven.
   1965   *The Kapauku Papuans of West New Guinea.* Holt, Rinehart and Winston, New York.

Puleston, Dennis E.
   1976   The Rio Hondo Project, Northern Belize. *Katunob* 9:29.
   1977   Experiments in Prehistoric Raised Field Agriculture: Learning from the Past. *Journal of Belizean Affairs* 5:36–43.
   1978   Terracing, Raised Fields, and Tree Cropping in the Maya Lowlands: A New Perspective on the Geography of Power. In *Pre-Hispanic Maya Agriculture*, edited by Peter D. Harrison and B. L. Turner II, pp. 225–45. University of New Mexico Press, Albuquerque.

Pyburn, Anne
   1987   Settlement Patterns at Nohmul, a Prehistoric Maya City in Northern Belize, Central America. *Mexicon* 9:110–14.
   1989   *Prehistoric Maya Community and Settlement at Nohmul, Belize.* BAR International Series 509. Oxford.

Redfield, Robert, and Alfonso Villa Rojas
   1934   *Chan Kom: A Maya Village.* University of Chicago Press, Chicago.

Reed, Erik K.
   1956   Types of Village Plan Layouts in the Southwest. In *Prehistoric Settlement Patterns in the New World*, edited by Gordon R. Willey, pp. 11–17. Viking Fund Publications in Anthropology, no. 23. New York.

Reyes Corts, M., and Joaquin Garcia-Barcena
   1979   Estratification en el Area de la Catedral. In *El Recinto Sagrado de México*, coordinated by C. Vega Sosa, pp. 16–28. Secretaria de Educación Pública–Instituto Nacional de Antropología e História, Mexico City.

Reynolds, S. E.
   1956   *Climatological Summary, New Mexico: Temperature 1849–1954.* Prepared

in cooperation with the New Mexico Interstate Stream Commission and the U.S. Department of Commerce, Weather Bureau. Technical Report no. 6. State of New Mexico, State Engineer Office, Santa Fe.

Riley, Carroll L.
1987   *The Frontier People: The Greater Southwest in the Protohistoric Period*. University of New Mexico Press, Albuquerque.

Riley, T. J.
1987   Ridged-Field Agriculture and the Mississippian Economic Pattern. In *Emergent Horticultural Economies of the Eastern Woodlands*, edited by William F. Keegan, pp. 295–304. Center for Archaeological Investigations Occasional Papers, no. 7. Southern Illinois University, Carbondale.

Roosevelt, A. C.
1980   *Parmana: Prehistoric Maize and Manioc Subsistence along the Amazon and Orinoco*. Academic Press, New York.

Rovner, Irwin
1983   Plant Opal Phytolith Analysis: Major Advances in Archaeobotanical Research. In *Advances in Archaeological Method and Theory*, vol. 6, edited by Michael B. Schiffer, pp. 225–66. Academic Press, New York.

Roys, Ralph L.
1931   *The Ethnobotany of the Maya*. Middle American Research Institute Publication no. 2. Tulane University, New Orleans.

Roys, Ralph L., France V. Scholes, and Eleanor B. Adams
1940   Report and Census of the Indians of Cozumel [1570]. *Contributions to American Anthropology and History* 6(30): 1–30.

1959   Census and Inspection of the Town of Pencuyut, Yucatan [1593], edited by Diego Garcia de Palacio, Oidor of the Audiencia of Guatemala. *Ethnohistory* 6:195–225.

Ruthenberg, H.
1971   *Farming Systems in the Tropics*. Clarendon Press, Oxford.

Sack, R. D.
1972   Geography, Geometry, and Explanation. *Annals of the Association of American Geographers* 62:61–78.

Sanders, William T.
1957   Tierra y Agua. Ph.D. diss., Department of Anthropology, Harvard University, Cambridge.

1965   *The Cultural Ecology of the Teotihuacan Valley*. Department of Sociology and Anthropology, Pennsylvania State University, University Park.

REFERENCES

1970a Comments to Contemporary Settlement Patterns: The Cerro Gordo–North Slope and Upper Valley Areas. In *The Natural Environment, Contemporary Occupation, and 16th Century Population of the Valley. The Teotihuacan Valley Project Final Report,* vol. 1, edited by William T. Sanders, A. Kovar, Thomas H. Charlton, and Richard A. Diehl, pp. 237–39. Occasional Papers in Anthropology, no. 3. Department of Anthropology, Pennsylvania State University, University Park.

1970b The Population of the Teotihuacan Valley, the Basin of Mexico and the Central Mexican Symbiotic Region in the 16th Century. In *The Natural Environment, Contemporary Occupation, and 16th Century Population of the Valley. The Teotihuacan Valley Project Final Report,* vol. 1, edited by William T. Sanders, A. Kovar, Thomas H. Charlton, and Richard A. Diehl. Occasional Papers in Anthropology, no. 3. Department of Anthropology, Pennsylvania State University, University Park.

1976 The Agricultural History of the Basin of Mexico. In *The Valley of Mexico,* edited by Eric R. Wolf, pp. 101–59. University of New Mexico Press, Albuquerque.

1981 Classic Maya Settlement Patterns and Ethnographic Analogy. In *Lowland Maya Settlement Patterns,* edited by Wendy Ashmore, pp. 351–69. School of American Research and University of New Mexico Press, Albuquerque.

Sanders, William T., A. Kovar, Thomas H. Charlton, and Richard A. Diehl (editors)
1970 *The Natural Environment, Contemporary Occupation, and 16th Century Population of the Valley. The Teotihuacan Valley Project Final Report,* vol. 1. Occasional Papers in Anthropology, no. 3. Department of Anthropology, Pennsylvania State University, University Park.

Sanders, William T., Jeffrey R. Parsons, and Robert S. Santley
1979 *The Basin of Mexico: Ecological Processes in the Evolution of a Civilization.* Academic Press, New York.

Sanders, William T., and B. J. Price
1968 *Mesoamerica: The Evolution of a Civilization.* Random House, New York.

Santley, Robert S.
1977 Intra-Site Settlement Patterns at Loma Torremote and Their Relationship to Formative Prehistory in the Cuauhtitlan Region, State of Mexico. Ph.D. diss., Department of Anthropology, Pennsylvania State University, University Park.

1987 A Consideration of the Olmec Phenomenon in the Tuxtlas: Middle Formative Settlement Pattern, Land Use, and Refuse Disposal at

Matacapan, Veracruz, Mexico. Paper presented at the 52d Annual Meeting of the Society for American Archaeology, Toronto.

Santley, Robert S., Thomas W. Killion, and Mark L. Lycett
1986   On the Maya Collapse. *Journal of Anthropological Research* 42:123–59.

Santley, Robert S., Ponciano Ortiz, P. J. Arnold III, B. A. Hall, Veronica Kann, J. M. Kerley, R. R. Kneebone, David Mora, R. Olivares, C. Parra, Christopher A. Pool, Z. Salazar, M. P. Smyth, and Clair Yarborough
1985a   *Final Field Report, Matacapan Project: 1984 Season.* Final report to the Instituto Nacional de Antropologia e História and the National Science Foundation. Department of Anthropology, University of New Mexico, Albuquerque.

Santley, Robert S., Ponciano Ortiz, P. J. Arnold III, R. R. Kneebone, M. P. Smyth, and J. M. Kerley
1985b   Reporte Final de Campo, Proyecto Matacapan: Temporada 1983. *Cuadernos del Museo de Antropologia de Universidad Veracruzana* 4:3–98.

Santley, Robert S., Ponciano Ortiz, Thomas W. Killion, P. J. Arnold III, and J. M. Kerley
1984   *Final Field Report of the Matacapan Archaeological Project: The 1982 Season.* Research Papers Series, no. 15. Latin American Institute, University of New Mexico, Albuquerque.

Santley, Robert S., Ponciano Ortiz, and Christopher A. Pool
1987   Recent Archaeological Research at Matacapan, Veracruz: A Summary of the Results of the 1982 to 1986 Field Seasons. *Mexicon* 9:41–48.

Sauer, Carl O.
1952   *Agricultural Origins and Dispersals.* American Geographical Society, New York.
1963   *Land and Life: A Selection from the Writings of C. O. Sauer,* edited by J. Leighly. University of California Press, Berkeley and Los Angeles.

Sayles, Edwin B.
1963   *An Archaeological Study of Chihuahua, Mexico.* Medallion Papers, no. 22. Gila Pueblo, Globe, Arizona.

Scarborough, Vernon
1983   Raised Field Detection at Cerros, Northern Belize. In *Drained Field Agriculture in Central and South America,* edited by J. P. Darch, pp. 123–36. BAR International Series 189. Oxford.

Schiffer, Michael B.
1976   *Behavioral Archaeology.* Academic Press, New York.
1977   Toward a Unified Science of the Cultural Past. In *Research Strategies in*

Historical Archaeology, edited by Stanley South, pp. 13–50. Academic Press, New York.

Schoenwetter, James, and Alfred E. Dittert, Jr.
   1968   An Ecological Interpretation of Anasazi Settlement Patterns. In *Anthropological Archaeology in the Americas,* edited by Betty J. Meggers, pp. 41–66. Anthropological Society of Washington, Washington, D.C.

Scholes, France V., and Ralph L. Roys
   1968   *The Maya Chontal Indians of Acalan-Tixchel.* University of Oklahoma Press, Norman.

Schoolcraft, H. R.
   1860   *Archives of Aboriginal Knowledge.* Vol. 1. J. P. Lippincott, Philadelphia.

Seele, Enno
   1973   Restos de Milpas y Poblaciones Prehispanicas cerca de San Buenaventura Neatlican, Puebla. *Communicaciones, Fundación Alemana por la Investigación Científica* 7:78–83.

Shafer, Harry J.
   1983   The Lithic Artifacts of the Pulltrouser Area: Settlements and Fields. In *Pulltrouser Swamp: Ancient Maya Habitat, Agriculture, and Settlement in Northern Belize,* edited by B. L. Turner II and Peter D. Harrison, pp. 212–45. University of Texas Press, Austin.

Shafer, Harry J., and Thomas R. Hester
   1983   Ancient Maya Chert Workshops in Northern Belize, Central America. *American Antiquity* 48:519–43.

   1986   An Ancient Maya Hafted Stone Tool from Northern Belize. Working Papers in Archaeology, no. 3. Center for Archaeological Research, University of Texas, San Antonio.

Sharer, Robert J.
   1974   The Prehistory of the Southeastern Maya Periphery. *Current Anthropology* 15(2): 165–87.

Sharer, Robert J. (editor)
   1978   *The Prehistory of Chalchuapa, El Salvador.* University of Pennsylvania Press, Philadelphia.

Sheets, Payson D.
   1971   An Ancient Natural Disaster. *Expedition* 14(1): 24–31.

   1976   Ilopango Volcano and the Maya Protoclassic. University Museum Studies, no. 9. Southern Illinois University, Carbondale.

   1982   Prehistoric Agricultural Systems in El Salvador. In *Maya Subsistence:*

Studies in Memory of Dennis E. Puleston, edited by Kent V. Flannery, pp. 99–118. Academic Press, New York.

1989   Introduction. In *1989 Archaeological Investigations at the Cerén Site, El Salvador: A Preliminary Report*, edited by Payson D. Sheets and Brian R. McKee, pp. 1–5. Department of Anthropology, University of Colorado, Boulder.

Sheets, Payson D. (editor)
1983   *Archaeology and Volcanism in Central America: The Zapotitán Valley of El Salvador.* University of Texas Press, Austin.

Sheets, Payson D., and Brian R. McKee (editors)
1989   *1989 Archaeological Investigations at the Cerén Site, El Salvador: A Preliminary Report.* Department of Anthropology, University of Colorado, Boulder.

Short, Susan K.
1983   Pollen from 1978 Research in the Zapotitán Valley. In *Archaeology and Volcanism in Central America: The Zapotitán Valley of El Salvador*, edited by Payson D. Sheets, Appendix II, pp. 300–302. University of Texas Press, Austin.

Siemens, Alfred H.
1982   Pre-Hispanic Use of the Wetlands of Northern Belize. In *Maya Subsistence: Studies in Memory of Dennis E. Puleston*, edited by Kent V. Flannery, pp. 205–25. Academic Press, New York.

Siemens, Alfred H., R. J. Hebda, Mario Navarrete Hernandez, D. R. Piperno, Julie K. Stein, and M. G. Zola Baez
1988   Evidence for a Cultivar and a Chronology from Patterned Wetlands in Central Veracruz, Mexico. *Science* 242:105–07.

Siemens, Alfred H., and Dennis E. Puleston
1972   Ridged Fields and Associated Features in Southern Campeche: New Perspectives on the Lowland Maya. *American Antiquity* 37:228–39.

Siliceo Pauer, Paul
1922   Condiciones físicos-biológicas. In *La población del valle de Teotihuacan*, vol. 4, edited by Manuel Gamio, pp. 147–201. Instituto Nacional Indigenista, Mexico, D. F.

Simeon, R.
1984   *Diccionario de la Lengua Nahuatl o Mexicana.* Siglo Ventiuno, Mexico City.

Simoons, Frederick J.
1965   Two Ethiopian Gardens. *Landscape* 14:15–26.

REFERENCES

Skinner, S. Alan
1965 A Survey of Field Houses at Sapawe, North Central New Mexico. *Southwestern Lore* 31:18–24.

Smith, Bruce D.
1987 The Independent Domestication of Indigenous Seed-bearing Plants in Eastern North America. In *Emergent Horticultural Economies of the Eastern Woodlands*, edited by William F. Keegan, pp. 3–47. Center for Archeological Investigations Occasional Papers, no. 7. Southern Illinois University, Carbondale.

Smith, C. E.
1980 Plant Remains from the Chiriqui Sites and Ancient Vegetational Patterns. In *Adaptive Radiations in Prehistoric Panama*, edited by Olga F. Linares and A. J. Ranere, pp. 151–74. Peabody Museum Monographs, Harvard University, no. 5. Cambridge.

Smith, M. Estellie
1967 Aspects of Social Control Among the Taos Indians. Ph.D. diss., Department of Anthropology, State University of New York, Buffalo.

Smith, Philip E. L., and T. Cuyler Young, Jr.
1972 The Evolution of Early Agriculture and Culture in Greater Mesopotamia: A Trial Model. In *Population Growth: Anthropological Implications*, edited by Brian Spooner, pp. 1–59. MIT Press, Cambridge.

Stadelman, Raymond
1940 *Maize Cultivation in Northwestern Guatemala.* Contributions to American Anthropology and History, no. 33. Carnegie Institution of Washington Publication no. 523. Washington, D.C.

Stark, Barbara
1987 Survey Methods in the Proyecto Arqueologico La Mixtequilla: A Comparative Discussion. Paper prepared for Balance y Perspectivas de la Antropología en Veracruz, conference, November 24–27, Jalapa, Mexico.

Steen, Charles R.
1977 *Pajarito Plateau Archaeological Survey and Excavations.* Los Alamos Scientific Laboratories Report no. LASL-77-4. Los Alamos.

Steggerda, Morris
1941 *Maya Indians of Yucatan.* Carnegie Institution of Washington Publication no. 531. Washington, D.C.

Stevenson, Matilda (Coxe)
1904 *The Zuni Indians: Their Mythology, Esoteric Fraternities, and Ceremonies.*

Twenty-third Annual Report of the Bureau of American Ethnology for the Years 1901–1902, pp. 3–634. Washington, D.C.

Stewart, Guy R.
1940   Conservation in Pueblo Agriculture II: Present-Day Flood Water Irrigation. *Scientific Monthly* 51:329–40.

Stewart, Guy R., and Maurice Donnelly
1943   Soil and Water Economy in the Pueblo Southwest. *Scientific Monthly* 56:31–44, 134–44.

Swadesh, Francis Leon
1974   *Los Primeros Pobladores: Hispanic Americans of the Ute Frontier.* University of Notre Dame Press, Notre Dame, Indiana.

Szymanski, Richard, and J. A. Agnew
1981   *Order and Skepticism: Human Geography and the Dialectic of Science.* Association of American Geographers, Washington, D.C.

Taschek, Jennifer T., and Joseph W. Ball
1986   Guerra: A Late Classic Suburban Paraje of Buenavista del Cayo, Belize. Paper presented at the 51st Annual Meeting of the Society for American Archaeology, New Orleans.

1987   Regal-Ritual Residences and Administrative Hubs: Differential Structure and Function among the Major Centers of the Upper Belize Valley. Paper presented at the 52d Annual Meeting of the Society for American Archaeology, Toronto.

Thomas, Prentice M., Jr.
1981   *Prehistoric Maya Settlement Patterns at Becan, Campeche, Mexico.* Middle American Research Institute Publication no. 45. Tulane University, New Orleans.

Thompson, J. Eric S.
1966   *The Rise and Fall of Maya Civilization.* University of Oklahoma Press, Norman.

Titiev, Mischa
1944   *Old Oraibi: A Study of the Hopi Indians of Third Mesa.* Papers of the Peabody Museum of Archaeology and Ethnography, vol. 22, no. 1. Harvard University, Cambridge.

Tjaden, Rex L.
1979   Bordered Garden Plots and Field Houses Near Sapawe, North Central New Mexico: A Dry Farming Strategy. Master's thesis, Department of Anthropology, Arizona State University, Tempe.

REFERENCES

Tozzer, Alfred M. (editor and translator)
   1941   *Landa's Relación de las Cosas de Yucatán*. Papers of the Peabody Museum of Archaeology and Ethnology, vol. 18. Harvard University, Cambridge.

Turner, B. L. II
   1974   Prehistoric Intensive Agriculture in the Maya Lowlands: New Evidence from the Rio Bec Region. Ph.D. diss., Department of Geography, University of Wisconsin, Madison.
   1978   The Development and Demise of the Swidden Thesis of Maya Agriculture. In *Pre-Hispanic Maya Agriculture*, edited by Peter D. Harrison and B. L. Turner II, pp. 13–22. University of New Mexico Press, Albuquerque.
   1983a Constructional Inputs for Major Agrosystems of the Ancient Maya. In *Drained Field Agriculture in Central and South America*, edited by J. P. Darch, pp. 11–26. BAR International Series 189. Oxford.
   1983b The Excavation of Raised and Channelized Fields at Pulltrouser Swamp. In *Pulltrouser Swamp: Ancient Maya Habitat, Agriculture, and Settlement in Northern Belize*, edited by B. L. Turner II and Peter D. Harrison, pp. 30–51. University of Texas Press, Austin.
   1983c *Once Beneath the Forest: Prehistoric Terracing in the Rio Bec Region of the Maya Lowlands*. Dellplain Latin American Studies, no. 13. Westview Press, Boulder.
   1989   The Specialist-Synthesis Approach to the Revival of Geography: The Case of Cultural Ecology. *Annals of the Association of American Geographers* 79.

Turner, B. L. II, and S. B. Brush
   1987   *Comparative Farming Systems*. Guilford Press, New York.

Turner, B. L. II, and William E. Doolittle
   1978   The Concept and Measure of Agricultural Intensity. *Professional Geographer* 30:297–301.

Turner, B. L. II, R. Q. Hanham, and Anthony V. Portararo
   1977   Population Pressure and Agricultural Intensity. *Annals of the Association of American Geographers* 67:384–96.

Turner, B. L. II, and Peter D. Harrison (editors)
   1983   *Pulltrouser Swamp: Ancient Maya Habitat, Agriculture, and Settlement in Northern Belize*. University of Texas Press, Austin.

Turner, B. L. II, and Charles H. Miksicek
   1984   Economic Plant Species Associated with Prehistoric Agriculture in the Maya Lowlands. *Economic Botany* 38:179–193.

Turney, Omar A.
1929   *Prehistoric Irrigation in Arizona*. Arizona State Historian, Phoenix.

Udo, R. K.
1965   Disintegration of Nucleated Settlement in Eastern Nigeria. *Geographical Review* 55:53–66.

Underhill, Ruth M.
1946   *Work a Day Life of the Pueblos*. Indian Life and Customs, no. 4. Phoenix Indian School, USDI Bureau of Indian Affairs, Phoenix.

Vega Sosa, C. (coordinator)
1979   *El Recinto Sagrado de Mexico*. Secretaría de Educación Pública–Instituto Nacional de Antropología e História, Mexico City.

Vivian, R. Gwinn
1970   An Inquiry into Prehistoric Social Organization in Chaco Canyon, New Mexico. In *Reconstructing Prehistoric Pueblo Societies*, edited by William A. Longacre, pp. 59–83. University of New Mexico Press, Albuquerque.

1974   Conservation and Diversion: Water Control Systems in the Anasazi Southwest. In *Irrigation's Impact on Society*, edited by Theodore E. Downing and McGuire Gibson, pp. 95–112. Anthropological Papers of the University of Arizona, no. 25. Tucson.

1984   Agricultural and Social Adjustments to Changing Environments in the Chaco Basin. In *Prehistoric Agricultural Strategies in the Southwest*, edited by Suzanne K. Fish and Paul R. Fish, pp. 243–57. Anthropological Research Papers, no. 33. Arizona State University, Tempe.

Vogt, Evon Z.
1969   *Zinacantan: A Maya Community in the Highlands of Chiapas*. Harvard University Press, Cambridge.

Wagner, Phillip L.
1958   Nicoya: A Cultural Geography. *University of California Publications in Geography* 12:195–258.

Walling, Stanley L., and Leah Minc
n.d.   Excavations at Tibaat, Pulltrouser Swamp. Ms. in possession of authors.

Warren, A. H.
1979   Ceramic Studies in the Cochiti Reservoir, 1976–1977. In *Archaeological Investigations in Cochiti Reservoir, New Mexico, Vol. 3: 1976–1977 Field Seasons*, edited by J. V. Biella, pp. 27–39. Office of Contract Archeology, University of New Mexico, Albuquerque.

REFERENCES

Wauchope, Robert
   1938   *Modern Maya Houses: A Study of Their Archaeological Significance.* Carnegie Institution of Washington Publication no. 502. Washington, D.C.

Weatherwax, Paul
   1954   *Indian Corn in Old America.* Macmillan, New York.

Wendorf, Fred
   1953   Excavations at Te'ewi. In *Salvage Archaeology in the Chama Valley, New Mexico,* assembled by Fred Wendorf, pp. 34–93. Monographs of the School of American Research, no. 17. Santa Fe.

Wendorf, Fred, and Erik K. Reed
   1955   An Alternative Reconstruction of Northern Rio Grande Prehistory. *El Palacio* 62:131–73.

Wenke, R. J.
   1984   *Patterns in Prehistory: Humankind's First Three Million Years.* Oxford University Press, New York.

West, R. C.
   1947   *Cultural Geography of the Modern Tarascan Area.* Smithsonian Institution, Institute of Social Anthropology Publication no. 7. Greenwood Press, Westport, Connecticut.

Whiting, Alfred F.
   1939   *Ethnobotany of the Hopi.* Bulletin of the Museum of Northern Arizona, no. 15. Flagstaff.

Whiting, Alfred F., Volney H. Jones, and Edmund Nequatewa
   1935   Interviews from the Hopi Crop Survey. Ms. on file, Museum of Anthropology, University of Michigan, Ann Arbor.

Wilhelm, Gene, Jr.
   1975   Dooryard Gardens and Gardening in the Black Community of Brushy, Texas. *Geographical Review* 65:73–92.

Wilk, Richard R.
   1983   Households in Process: Agricultural Change and Domestic Transition among the Kekchi Maya of Belize. In *Households: Comparative and Historical Studies of the Domestic Group,* edited by R. M. Netting, Richard R. Wilk, and E. J. Arnould, pp. 217–44. University of California Press, Berkeley and Los Angeles.

Wilken, Gene C.
   1971   Food-Producing Systems Available to the Ancient Maya. *American Antiquity* 36:432–48.

REFERENCES

1987 *Good Farmers: Traditional Agricultural Resource Management in Mexico and Central America*. University of California Press, Berkeley and Los Angeles.

Williams, B. J.
1989 Contact Period Rural Overpopulation in the Basin of Mexico: Carrying-Capacity Models Tested with Documentary Data. *American Antiquity* 54:715–32.

Wiseman, Frederick M.
1983 Analysis of Pollen from the Fields at Pulltrouser Swamp. In *Pulltrouser Swamp: Ancient Maya Habitat, Agriculture, and Settlement in Northern Belize*, edited by B. L. Turner II and Peter D. Harrison, pp. 105–19. University of Texas Press, Austin.

Wittfogel, Karl A., and Ester S. Goldfrank
1943 Some Aspects of Pueblo Mythology and Society. *Journal of American Folklore* 56:17–30.

Wobst, H. Martin
1978 The Archaeo-Ethnology of Hunter-Gatherers or the Tyranny of the Ethnographic Record in Archaeology. *American Antiquity* 43:303–09.

Wolf, Eric R.
1959 *Sons of the Shaking Earth*. University of Chicago Press, Chicago.

Woodbury, Richard B.
1961 *Prehistoric Agriculture at Point of Pines, Arizona*. Memoirs of the Society for American Archaeology, no. 17. Salt Lake City.

1966 Evidence of Prehistoric Farming in the Vicinity of Picuris, New Mexico. Smithsonian Institution, Washington, D.C. Ms. on file, Laboratory of Anthropology, Museum of New Mexico, Santa Fe.

Woodbury, Richard B., and Ezra B. W. Zubrow
1979 Agricultural Beginnings, 2000 B.C.–A.D. 500. In *Handbook of North American Indians*, vol. 9, edited by William C. Sturtevant, pp. 43–60. Smithsonian Institution, Washington, D.C.

Woosley, Anne I.
1980 Agricultural Diversity in the Prehistoric Southwest. *Kiva* 45:317–35.

1986 Puebloan Prehistory of the Northern Rio Grande: Settlement, Population, Subsistence. *Kiva* 51:143–64.

Woot-Tsuen, W. L., with the cooperation of M. Flores
1961 *Food Composition Table for Use in Latin America*. Interdepartmental Committee on Nutrition for National Defense and the Institute of

REFERENCES

Nutrition of Central America and Panama, Bethesda, Maryland, and Guatemala City.

Wozniak, Frank E.
1986 *Irrigation in the Rio Grande Valley, New Mexico: An Annotated Bibliography.* Prepared for New Mexico Historic Preservation Division, Santa Fe, under an intergovernmental agreement with USDI Bureau of Reclamation, Southwest Regional Office, Amarillo, Contract no. BOR-86-1.

Wright, A. C. S., D. H. Romney, R. H. Arbuckle, and V. E. Vial
1959 *Land in British Honduras: Report of the British Land Use Survey Team.* Colonial Research Publications, no. 24. Her Majesty's Stationery Office, London.

Yarnell, Richard A.
1965 Implications of Distinctive Flora on Pueblo Ruins. *American Anthropologist* 67:662–73.

Zier, Christian J.
1980 A Classic-Period Maya Agricultural Field in Western El Salvador. *Journal of Field Archaeology* 7:65–74.

1981 *A Functional Analysis of Late Classic Period Maya Settlements in the Zapotitán Valley, El Salvador.* Ph.D. diss., Department of Anthropology, University of Colorado, Boulder. University Microfilms, Ann Arbor.

1983 The Cerén Site: A Classic Period Maya Residence and Agricultural Field in the Zapotitán Valley of El Salvador. In *Archaeology and Volcanism in Central America: The Zapotitán Valley of El Salvador,* edited by Payson D. Sheets, pp. 119–43. University of Texas Press, Austin.

Contributors

Kurt F. Anschuetz has participated in archaeological research projects in the central Andes, the Valley of Mexico, and the northern Southwest of the United States. He became interested in the agricultural technologies and social organization of late prehistoric and early historic Pueblo Indians of the northern Rio Grande of New Mexico while working on his M.A. at the University of New Mexico (1984) and the Office of Archaeological Studies, Museum of New Mexico (1985–1987). His research also includes issues of Pueblo Indian ethnology, particularly uses and concepts of corn and water. He is currently a doctoral candidate at the Museum of Anthropology, University of Michigan, where he is preparing to return to the lower Rio Chama valley to continue his studies of fourteenth- and fifteenth-century puebloan peoples and their gravel-mulched field systems.

Joseph W. Ball, professor of anthropology at San Diego State University, received his Ph.D. from the University of Wisconsin at Madison in 1974. He is a specialist in ceramic typology, and his interests have centered on applying the principles of behavioral archaeology and contextual analysis to illuminating ancient Maya community structure and political organization. He has participated in fieldwork at Becan, Campeche, and Kaminaljuyu, Guatemala, and since 1981 has been co-

CONTRIBUTORS

director of the SDSU Mopan-Macal Triangle Archaeological Project in Belize. His publications include *The Archaeological Ceramics of Becan, Campeche, Mexico* (1977), *The Archaeological Ceramics of Chinkultic, Chiapas, Mexico* (1980), "The Rise of the Northern Maya Chiefdoms" (1977), "Teotihuacan's Fall and the Rise of the Itza" (1989), and, with Jennifer Taschek, "Late Classic Lowland Maya Political Organization and Central-Place Analysis" (1991).

William E. Doolittle, associate professor of geography at The University of Texas at Austin, received a Ph.D. from the University of Oklahoma in 1979. He is a specialist in the analysis of dry-land agricultural systems, including prehistoric, historic, or small-scale traditional systems currently in use. He has carried out research in Mexico and the American Southwest with funding from the National Science Foundation, the National Endowment for the Humanities, and various private sources. His publications include numerous articles in major geographical and archaeological journals as well as two books—*Pre-Hispanic Occupance in the Valley of Sonora, Mexico: Archaeological Confirmation of Early Spanish Reports* (1988) and *Canal Irrigation in Prehistoric Mexico: The Sequence of Technological Change* (1990). He is currently writing a book on the agricultural landscapes of North America on the eve of European Contact.

Susan T. Evans is an archaeologist specializing in the Aztec period of the Teotihuacan Valley. This research focus involves settlement and agricultural studies (such as described in the paper in this volume) as well as testing ethnohistoric models of sociopolitical organization and ideological orientation. She edited and contributed to *Excavations at Cihuatecpan* (1988), and more recently has coauthored *Out of the Past: An Introduction to Archaeology* (1992) with Pennsylvania State University colleagues David Webster and William Sanders. She received her Ph.D. from Pennsylvania State University in 1980.

Richalene G. Kelsay is a senior environmental analyst for CALTRANS (California Department of Transportation) at the San Diego southern California regional office. Since leaving a first career as chief costume designer and script writer for the popular long-running television series, "Little House on the Prairie," she has participated in archaeological projects throughout southern California and in Belize. She currently is an M.A. candidate in the anthropology department at San Diego State University. Her research interests center around the archaeological recognition and interpretation of anthrosols on southern California collector-gatherer sites.

Thomas W. Killion received his Ph.D. in anthropology from the University of New Mexico in 1987. He presently holds a research associate position in the Department of Archaeology at Boston University and is an archaeologist working in the Repatriation Office at the Smithsonian Institution. He has conducted archaeological and ethnoarchaeological fieldwork in southern Veracruz, Mexico, and also is involved with ongoing archaeological research in Guatemala and Belize. His research focus is the emergence of agricultural systems in tropical lowland Mesoamerica and the subsequent role of agricultural production in the rise of lowland urban centers. Recent articles and research papers have examined the ethnoarchaeology of peasant agriculture in southern Veracruz, intensive archaeological survey at the prehispanic urban center of Sayil, Yucatan, Mexico, and the intersite survey of settlement, agricultural terraces, and wall systems in the Petexbatun region of Peten, Guatemala.

Timothy D. Maxwell is assistant director at the Museum of New Mexico, Office of Archaeological Studies. A doctoral candidate at the University of New Mexico, he is currently conducting a comparative study of prehistoric farming methods in the Zuni and lower Rio Chama regions of New Mexico. His special interest is the intermediary relationship that farming technology plays between environment and agricultural productivity.

Patricia A. McAnany, assistant professor of archaeology at Boston University, received a Ph.D. in Anthropology from the University of New Mexico in 1986. Since 1981 she has carried out archaeological field research at several locales of Maya settlement including Pulltrouser Swamp, Colha, and Sayil. Her research interests include economic organization, particularly of stone tool production and distribution, and the agricultural landscapes of Classic Maya society as they relate to settlement structure. On these topics she has published several articles and, with Barry L. Isaac, edited *Prehistoric Maya Economies of Belize*. More recently, she has initiated an investigation into the formative roots of ancestor veneration and lineage organization at K'axob, Belize.

William T. Sanders is Evan Pugh Professor of Anthropology at Pennsylvania State University. He received his Ph.D. degree from Harvard University in 1957 and is a member of the National Academy of Sciences. His primary interest is the evolution of complex society, and he approaches this question with an ecological perspective. He has worked in many areas in Mesoamerica, but his major research projects have been in the Valley of Mexico, Kaminaljuyu, Guatemala, and Copan, Honduras. He is co-author of several books, a large number of monographs, book chapters, and journal articles.

Robert S. Santley is professor of anthropology at the University of New Mexico. He received his Ph.D. in anthropology from the Pennsylvania State University in 1977. He has conducted fieldwork in Pennsylvania, Guatemala, Central Mexico, and on the south gulf coast of Mexico. From 1982 to 1987, he acted as director of the Matacapan Project, and most recently he has been conducting a program of archaeological survey in the Tuxtlas mountains. He is author of numerous articles and book chapters and coauthor, with William T. Sanders and Jeffrey R. Parsons, of *The Basin of Mexico: Ecological Processes in the Evolution of a Civilization* (1979).

B. L. Turner II, director, George Perkins Marsh Institute and a professor of geography, Clark University, received a Ph.D. from the University of Wisconsin-Madison in 1974. He has worked extensively on ancient Maya environment, food production, and population based on fieldwork in the Yucatan peninsular region, Belize, Guatemala, and Honduras. His books and edited volumes include: *Pre-Hispanic Maya Agriculture,* with P. D. Harrison (1978), *Once Beneath the Forest* (1983), and *Pulltrouser Swamp: Ancient Maya Habitat, Agriculture, and Settlement in Northern Belize,* with P. D. Harrison (1983), *Comparative Farming Systems,* with S. B. Brush (1987), *The Earth as Transformed by Human Action,* with others (1990). He currently is working on agricultural and global land-use change and on the environmental impacts of the Columbian Encounter.

Christian J. Zier is owner and president of Centennial Archaeology, Inc., a consulting firm based in Fort Collins, Colorado. He received a Ph.D. in anthropology from the University of Colorado-Boulder in 1981. He has worked on archaeological and paleontological projects in the western and central United States, North Africa, and Central America, and has published numerous articles as well as several chapters in books and monographs. Interests include hunter-gatherer cultural ecology and settlement, agricultural adaptations in marginal environments, and archaeological site formation and structure. His present research emphasis is Archaic and Woodland adaptations of the Colorado Piedmont area of eastern Colorado.

Index

Agricultural form and function, 15–19, 268–69
Agricultural intensification, 10–12, 23, 187–88, 232
   defined, 185, 187, 271
   goal of, 272
   in Pulltrouser Swamp, 184, 187–90
Agricultural strategies (practices), 2–3, 4–8. *See also* Agricultural intensification; Horticultural strategies (practices)
   of Anasazi culture, 45, 67–68
   during Aztec period, 93–96
   at the Cerén site, 224, 226–27, 230–32
   extensive, defined, 13 (n. 1)
   and garden-residence site structure, 131–32, 135 (table 6-2), 136–37
   monocropping, 16, 23, 267
Agricultural tasks. *See also* Horticultural strategies (practices)
   field (plot) maintenance, 201
      compared to field preparation, 190, 191 (fig. 8-2), 192 (fig. 8-3), 193
   labor demands of, 7–8, 99–100, 194–95
Agricultural tools. *See also* Bifaces, oval
   types and contexts of, at Pulltrouser Swamp, 202, 203 (fig. 8-5), 204–5, 207
Agriculture
   compared to horticulture, 267
   defined, 13 (n. 3)
Agrotechnologies. *See* Agricultural strategies (practices)
Aguamiel (fresh maguey sap), 97, 106–7, 110. *See also* Maguey: importance of
Akchin field, 48 (fig. 3-3)
   defined, 46–47

Alluvial plain, in Teotihuacan Valley, 97, 100, 102
Altepetlalli (city-state's land), 100, 102
Anasazi culture, 36
   agricultural systems of, 67–68
   and gravel-mulched fields, 57–58
   prehistory of, 54–55, 57
Aztec period, 94 (fig. 5-1), 96 (fig. 5-2), 102, 103
   agricultural practices during, 93–95, 97, 99–100
   defined, 114 (n. 1)
   landholding categories in, 100, 101 (table 5-1), 102

Bifaces, oval. *See also* Agricultural tools
   debitage from, 203 (fig. 8-5), 209, 210 (table 8-2), 211
   described, 202
   fragments of, analyzed, 206 (fig. 8-6), 207, 208 (table 8-1), 209
   methods of hafting, 202, 204, 280
Buenavista del Cayo site, 235, 245. *See also* Soil phosphate levels
   described, 244
   soil sample transects at, 256 (fig. 10-6)
   vacant terrain tests at, 255
Buenavista project, 234, 235
   methodology, 237, 239
   study area, 236 (fig. 10-1)

Calmilli, 96, 100, 275. *See also* Gardens; House-lot (kitchen) gardens

INDEX

demarcation of, 267–68
  as horticultural strategy, 105–6
  as house maize-fields, 18, 19
  in maguey production, 109–10
Calpullalli (peasants), 100, 102
Cerén site, the, 227–28, 229 (fig. 9-5)
  agricultural features of, 222, 224, 225 (fig. 9-3)
  agricultural practices at, 230–32
    effect of *tierra blanca* on, 224, 226–27
  conclusions about, project, 232–33
  critique of, project, 277–78
  described, 221–22, 223 (fig. 9-2)
Cerro San Lucas, 97, 98 (fig. 5-3)
  effect of Spanish conquest on, 102, 103
  horticultural strategies around, 103, 105–6
  maguey farming in, 107, 108 (table 5-2), 110–11 (*see also* Maguey)
Cerros de trincheras (hillside terraces), 74, 75
Chi Ak'al, 197, 199 (fig. 8-4)
Chinampas, 22, 96, 217, 272
Chontalpa, as example of Mesoamerican settlement agriculture, 23–26
Cihuatecpan, 98 (fig. 5-3)
  climate of, 103–4
  effect of Spanish conquest on, 102, 103
  historical background of, 92–95, 97, 99–100, 102, 103
  importance of maguey for, 92–93, 99, 111–12 (*see also* Maguey: as nutritional supplement)
Coa (farming implement), 188, 194, 196
Cochineal, 112–13
Codex Mendoza, tribute requirements of, 102, 106, 112, 114–15 (n. 4)
Compact village, defined, 21–22
Cuchumatan highlands, described, 26–27

Debitage. *See* Bifaces, oval
Dry-farm field, defined, 45

Ejidatarios, 24–25
Ethnographic data (model)
  for agricultural tasks, 190–93
  for agricultural tools, 193–96
  for the Cerén site, 227–28, 229 (fig. 9-5), 230
  in estimating maguey production, 107, 109
  in Gran Chichimeca project, 70
  limitations of, 37, 38–39, 51–53
  at Matacapan, 152–54
  methodologies applied to, 273–74
  of residence site structure, 120–23
    in the Tuxtlas region, 123–25 (*see also* House-lot model)
Exotic nutrients, 18, 30 (n. 1)

Fertilization, as horticultural strategy, 76–77
Fields. See also *Milpas; names of specific types of fields*
  at Hopi, 45, 46, 47, 48 (fig. 3-3), 49
  in lower Rio Chama Valley, 66, 67–68
  misidentified as houses, 57–58
  Southwestern, 39, 41
    classified, 44–45
  at Zuni, 47
Floodplain field, defined, 47, 49

Garden residences
  and agricultural technologies, 131–32, 135 (table 6-2), 136–37
  components of, 120, 122–23
  in different demographic contexts, 121–23
Gardens. See also *Calmilli;* House-lot (kitchen) gardens
  defined, 13 (n. 2), 265–66
  at Hopi, 40 (fig. 3-1), 41 (table 3-1), 42 (table 3-2), 43 (fig. 3-2), 44
  Southwestern, features of, 39, 41, 42, 44
  waffle, 41, 44, 58, 61
  at Zuni, 41, 42, 44, 61, 62 (fig. 3-5), 63 (fig. 3-6)
Gran Chichimeca, 71 (fig. 4-1)
  conclusions about, project, 89–91
  critique of, project, 281–82
  house-lot gardens in, 69–70, 82
    compared to *cerros de trincheras* (hillside terraces), 74, 76
    evidence of, postulated, 83, 84 (fig. 4-5), 89, 90
    features of, 78–79, 80 (fig. 4-3), 81 (fig. 4-4)
    horticultural techniques in, 76–78
    of the Pima Bajo, 70, 71, 72–74, 75 (fig. 4-2), 78, 79
    plants of, 72–74, 75 (fig. 4-2), 76, 281
    rock walls as evidence of, 85, 86 (fig. 4-6), 87 (fig. 4-7), 88 (fig. 4-8), 89 (fig. 4-9)
    of the Tarahumar, 70, 73, 74, 79
    of the Tepehuan, 70, 78, 79
Gravel-mulched fields
  in Anasazi culture, 57–58
  defined, 50–51
  in lower Rio Chama Valley
    excavation of, 62 (fig. 3-5), 63 (fig. 3-6)
    significance of, 61, 64–66
Guatemala, highland
  as example of Mesoamerican settlement agriculture, 26–28
Guerra site, 235, 245. *See also* Soil phosphate levels
  described, 239–41
  the, model, 241–43

331

INDEX

(Guerra site, *continued*)
  predictions for, 243-44
  radial tests at, 246 (fig. 10-3), 247-49
  soil sample transects at, 238 (fig. 10-2)
  vacant terrain tests at, 249-50, 252, 254, 258

Histograms (Matacapan project), 160 (fig. 7-3), 162-63 (fig. 7-4), 165 (fig. 7-5)
Hohokam culture, features of, 36
Hopi, 38-39, 51-52
  fields at, 45, 46, 47, 48 (fig. 3-3), 49
  gardens at, 40 (fig. 3-1), 41 (table 3-1), 42 (table 3-2), 43
Horticultural strategies (practices), 73, 79. *See also* Agricultural strategies (practices); Agricultural tasks; Terracing
  around Cerro San Lucas, 103, 105-6
  fertilization, 76-77
  in Gran Chichimeca, 76-78
  intercropping, 16, 27, 77, 107, 109, 110
  irrigation, 3, 36-37, 52-53, 76-77, 228
  planting, 77
  rock rings as, 83, 84 (fig. 4-5), 85
  transplanting, 77-78
  weeding, 77, 193-96, 280
Horticulture, defined, 267-68
House-compound model (and toft-zone/intermediate area), 167-69
House-lot (kitchen) gardens, 7-8, 20, 23, 25, 29. *See also Calmilli*; Horticultural strategies (practices)
  compared to *cerros de trincheras* (hillside terraces), 74, 76
  evidence of
    postulated, 83, 84 (fig. 4-5), 89, 90
    rock walls as, 85, 86 (fig. 4-6), 87 (fig. 4-7), 88 (fig. 4-8), 89 (fig. 4-9)
  factors in development of, 17-19
  features of, 78-79, 80 (fig. 4-3), 81 (fig. 4-4)
  in highland Guatemala, 27 (see also *Rastrajo*)
  importance of, 14
    for archaeologists, 69-70
  of the Pima Bajo, 70, 71, 72-74, 75 (fig. 4-2), 78, 79
  plants of, 72-74, 75 (fig. 4-2), 76, 281
  spatial relationships of, 82
  of the Tarahumar, 70, 73, 74, 79
  of the Tepehuan, 70, 78, 79
House-lot model, 125 (fig. 6-2), 128, 242-43
  compared to house-compound model (intermediate area or toft zone), 167-69
  components of, 124, 130 (table 6-1), 131, 274
    clear area, 127-28, 167
    garden area, 129-31, 167
    intermediate area, 128-29, 167

  structural core, 126, 167
  usefulness of, 275-76
House lots *(solars)* in the Tuxtlas, 127, 168
  northern, 133 (fig. 6-3)
  southern, 134 (fig. 6-4)
Huatal (bush-regrowth parcels), 27, 31 (n. 2)

Ilopango Volcano, 219, 221
Infield-outfield agriculture, 6, 19, 27-28, 29-30, 267-70
Infields, 26, 27-28, 29, 275
  defined, 266-67
  demarcation of, 25, 267-68
  Olmec, 161, 164, 166
  at Pulltrouser Swamp, 199 (fig. 8-4), 200-201
  in the Tuxtlas, 132, 135 (table 6-2), 136, 153-54
Intercropping
  as horticultural strategy, 16, 27, 77
  in maguey production, 107, 109, 110
Irrigated field, defined, 50
Irrigation, as horticultural strategy, 3, 36-37, 52-53, 76-77, 228

Jagueys (reservoirs), 104, 110-11

Kapauku (of Irian Jaya), 192 (fig. 8-3), 193, 194
K'axob, 197, 198, 199 (fig. 8-4)
Kitchen gardens. *See* House-lot (kitchen) gardens
Kokeal, 197, 198, 199 (fig. 8-4)

Laguna Caldera, 221
Landholding categories, Aztec, 100, 101 (table 5-1), 102
Landscape capital (modification), 3, 271
Linear-border field, defined, 49-50
Llano (short fallow sod land), 27-28, 31 (n. 2)
Loma Torremote, 175-76, 183 (n. 5)

Maguey, 73, 105
  importance of, 92-93, 99, 111-12
  leaves *(pencas)*, 99, 112
  as nutritional supplement, 106-7, 108 (table 5-2), 109-11, 113
  tools for processing, 97, 99
Maguey terrace agriculture
  conclusions about, project, 113-14
Maize
  cultivation of, in highland Guatemala, 26, 27, 28
  evidence of, at the Cerén site, 231

332

INDEX

in house-lot gardens, 74, 75 (fig. 4-2), 76, 281
Matacapan (Killion), 138–40. *See also* House-lot model
   archaeological zone, 138 (fig. 6-5)
   conclusions about, project, 148–49
   critique of, project, 274–76
   survey results from, 141 (fig. 6-6), 142 (fig. 6-7), 143–44
   use of settlement space at, analyzed, 143–44, 145 (table 6-3), 146–48
Matacapan (Santley), 156–57 (fig. 7-2)
   settlement patterns at, 154–55, 158–59, 161, 164, 166
Matacapan Archaeological Project, 137–38
Maya, 26, 27, 28, 196–97
Mesas
   prehistoric settlements on, 85, 87, 89, 90 (fig. 4-10)
Methodologies
   Buenavista project, 237, 239
   critique of, 274–83
Mexico, Basin of, 19–20
   during Aztec period, 93, 94 (fig. 5-1), 95, 114 (n. 1)
   effect of Spanish conquest on, 102, 103
   as example of Mesoamerican settlement agriculture, 21–23
Milpas, 100, 109, 110. *See also* Fields
   in Cihuatecpan, 95, 102
   defined, 114 (n. 2)
   as part of horticultural strategy, 105–6
Monocropping, 16, 23, 267
Montaña (forest parcels), 27, 31 (n. 2)

Nopal, 23, 99, 105, 112–13

Olmec. *See also* Histograms; Matacapan (Santley)
   artifacts, 169, 171, 173, 183 (n. 4)
   ceramics (pottery), 155, 158–59, 171–75
   functions of, 164, 166
   conclusions about, project, 177–81
   land-use strategies, 150, 152, 159, 161, 164, 166, 177–79
   obsidian, 155, 172–73, 175–76, 178–79
   tecomates (spherical neckless jars), 164, 166, 169
Orchard, defined, 266
Orchard-garden, defined, 266
Outfields, 23, 27, 29
   defined, 266–67
   Olmec, 161, 164, 166
   at Pulltrouser Swamp, 200–201, 270, 279–80
   in the Tuxtlas, 132, 135 (table 6-2), 153–54

Pajonal (grassland), 27–28, 31 (n. 2)
Pech Titon, 197, 199 (fig. 8-4)
Pima Bajo
   horticultural techniques of the, 76–78
   house-lot gardens of the, 70, 71, 72–74, 75 (fig. 4-2), 78, 79
Planting, as horticultural strategy, 77
Population density, 15, 19–20, 57, 102, 103
   during Aztec period, 93, 95, 97
   effects of, on agricultural practices, 5, 6–7, 18, 31, 93–94, 99–100
Pulltrouser Swamp, 186 (fig. 8-1), 196–97, 199 (fig. 8-4), 270. *See also* Agricultural tools; Bifaces, oval
   agricultural history of, 198, 200–201
   conclusions about, project, 211–13
   critique of, project, 278–81
   raised fields at, 201, 280–81
Pulque (fermented maguey sap), 97, 107. *See also* Maguey: importance of

Raised fields, 3, 22, 230
   at the Cerén site
     described, 222, 224, 225 (fig. 9-3)
     purposes of, 224, 226–27
   defined, 271
   functions of, 217–18, 271–72, 284 (n. 6)
   at Pulltrouser Swamp, 201, 280–81
   in Valley of Sonora, 87, 89, 90 (fig. 4-10)
Rancherías, 17, 21, 87–89, 90, 281
Rastrajo (cornstalk land), 28, 31 (n. 2)
Ridge and furrow, 11, 271
Ridged fields. *See* Raised fields
Ring chromatography, 237, 259, 260 (n. 2)
Rio Chama Valley, lower, 37–38. *See also* Hopi; Zuni
   agricultural features in, 58–60
   Anasazi culture in, 54–55, 57
   conclusions about, project, 67–68
   critique of, project, 282–83
   gravel-mulched fields in, 50, 61, 62 (fig. 3-5), 63 (fig. 3-6), 64–66
   land-use patterns in, 67–68
   prehistoric sites in, 53–54, 55 (fig. 3-4), 56 (table 3-3)
Rock enclosures, as horticultural strategy, 79
Rock rings, as horticultural strategy, 83, 84 (fig. 4-5), 85
Rock walls, as evidence of house-lot gardens, 86 (fig. 4-6), 87 (fig. 4-7), 88 (fig. 4-8), 89 (fig. 4-9)

San Diego State University Prehistoric Lowland Maya Community Structure Project. *See* Buenavista project
San Martin, 155

333

INDEX

Scattered village, defined, 21
Seed beds, 77–78, 79, 85
Seepage field, defined, 45–46
Settlement agriculture, 1–2, 15–19, 29–30
  defined, 14
  examples of, in Mesoamerica
    Basin of Mexico, 21–23
    Chontalpa, Tabasco, 23–26
    highland Guatemala, 26–28
Settlement patterns
  in conquest period Mesoamerica, 19–21
  factors in, 17–19
  at Matacapan, 154–55, 158–59, 161, 164, 166
  in Teotihuacan Valley, 96 (fig. 5-2), 100, 115 (n. 5)
Settlement-subsistence systems, 35–36
Shipibo, 190, 191 (fig. 8-2)
Slope-wash field, defined, 46
Soil categories in Chontalpa, 24
Soil phosphate levels. *See also* Buenavista del Cayo site; Guerra site; Ring chromatography
  at Buenavista, 237, 239, 257 (fig. 10-7)
  conclusions about, project, 257–59
  critique of, project, 276–77
  at Guerra, 237, 239, 251 (fig. 10-4), 253 (fig. 10-5)
Soil water conservation, 61, 64–65
Solars (house lots) in the Tuxtlas, 127, 168
  northern, 133 (fig. 6-3)
  southern, 134 (fig. 6-4)
Sonora, Valley of, 83, 85–87, 89, 90 (fig. 4-10)
Southwest (United States). *See also* Hopi; Rio Chama Valley, lower; Zuni
  archaeological studies of, 35
  errors in, 52–53
  fields in, 39, 41 (see also *names of specific types of fields*)
    classified, 44–45
  gardens in, 39, 41, 42, 44
Staples, 15–16, 28, 29
Subsistence crops, 25–26, 27
Subsistence farmers, 5, 6, 16–17, 20
  defined, 15

Tarahumar
  horticultural techniques of the, 76–78
  house-lot gardens of the, 70, 73, 74, 79
Tecomates (spherical neckless jars), 164, 166, 169
Tenochtitlan, 93, 95
Tenochtitlan-Tlateloco, 20
Teotihuacan Barrio, 183 (n. 3)

described, 169
Early Formative agricultural surface at, 171–72
refuse (debris) densities in, 174 (fig. 7-7)
topography of, 170 (fig. 7-6)
Teotihuacan Valley, 21–22, 99, 103–5, 106. *See also* Maguey; Maguey terrace agriculture
  alluvial plain in, 97, 100, 102
  land use in, 92, 94, 95 (*see also* Terracing: in Teotihuacan Valley)
  settlement patterns in, 96 (fig. 5-2), 100, 115 (n. 5)
Teotihuacan Valley Project, 95, 103
Tepehuan
  horticultural techniques of the, 76–78
  house-lot gardens of the, 70, 78, 79
Tephra (ash). See *Tierra blanca*
Terrace field, defined, 49
Terracing, 3
  in Teotihuacan Valley, 92, 93, 94, 95, 99–100
    abandoned, 103
    around Cerro San Lucas, 103, 105–6
    as environmental solution, 104–5
Tibaat, 197, 198, 199 (fig. 8-4)
Tierra blanca (white earth), 219, 221, 223 (fig. 9-2), 227, 228
  effect of, on agricultural practices, 224, 226–27
  erosion of, 222, 224, 226, 230–31
Toft-zone (house-compound model), 167–69
Toltec period, 93, 95, 97
Transplanting, as horticultural strategy, 77–78
Tuxtlas, the, 124 (fig. 6-1), 151 (fig. 7-1)
  described, 123
  house lots (*solars*) in, 127, 168, 133 (fig. 6-3), 134 (fig. 6-4)
  land-use strategies in, 152–54, 164

Urban village, described, 21

Waffle gardens, 41, 44, 58, 61
Water-table field, defined, 46
Weeding, as horticultural strategy, 77, 193–96, 280

Yo Tumben, 197, 198, 199 (fig. 8-4)

Zapotitán Valley, 220 (fig. 9-1)
Zuni
  akchin fields at, 47
  gardens at, 41, 42, 44, 61, 62 (fig. 3-5), 63 (fig. 3-6)